I0472964

Secret Life of the Brewer's Yeast: A Microbiology Tale
by David Wooster, Ph.D.

~ ~ ~

Copyright © 2024 / by David Wooster, Ph.D.

~ ~ ~

This book is a work of fiction.   Certain historical personages do appear briefly, but they are not given fictitious roles or inflated speeches.   Other characters are either the product of the author's imagination or are used fictitiously.   Any resemblance to actual persons, living or dead, events or locales is entirely coincidental.

~ ~ ~

TABLE OF CONTENTS

FALL SEMESTER

PART 1: FALL SEMESTER

THE OTHER KIND OF DOCTOR

Almost everyone gets it wrong so I should explain that I'm not the kind of doctor that practices medicine but rather the "other kind". We're usually professors most Ph.D.'s, which is why I teach as well as do research for a living. To my students I'm Dr. Ketchum, one of a handful of microbiologists in this small town along the foothills of the Rocky Mountains, but to most everyone else I'm just plain Ben. Lots of folks get the two types confused, though, and as I often explain more out of desperation than anything else when someone still doesn't get it is that, if I were to practice medicine, I'd go to jail just like anybody else without a medical license. But what bothers me even more these days is when my family reminds me that I'm pushing 40 and still not married. Maybe it's just that I've just never found the time, I'm not sure, but if it is an excuse it's one I've been using for some time now.

And as a microbiologist what interests me most is figuring out all the different ways these clever little "bugs" have found to get along in the world. Extreme microbiology, you could say, is my specialty. We're all supposed to have one these days. Well, it turns out these overlooked creatures have had more than three billion years of experimenting with different strategies for survival and there seem to be an endless number of ways they've come across, too, almost as many as there are microbes.[1]

On the other hand, we humans are the relative newcomers, an extensive collection of cells – around ten trillion of them – all grouped together and working side-by-side doing different jobs in vast colonies we call "tissues" and so we live our lives on a grand scale compared to single-celled life forms like bacteria and as a consequence tend to take for granted that rain, or wind, or most anything else that Mother Nature can throw at us on a regular basis doesn't have any real effect, except for perhaps having to pop open an umbrella once in a while.[i] Our size and complexity for the most part renders us indestructible to these everyday forces. But for microbes life can be a horse of a different color.

Changing conditions, maybe it's something as simple as a mud puddle drying up in the heat of the midday sun, or perhaps the danger lies closer to home, in the form of the sticky foot of some passing insect, either of which can spell the difference between life and death to these economical little beings and unlike us with our specialized tissues – muscles, bone, and skin

---

[1] And many are good at hitching a ride. About 10% of the weight you carry around every day is bacteria. Each of us has around 200 trillion microbes living in and on our bodies.

for instance – microbes are for the most part stuck where they are, without the elaborate defenses against nature. Well, most of them are anyway, wh ch brings me to my most recent setback.

The anthrax grant I had applied for, the one I had been counting on to fund my research for the next several years, was turned down by the National Institutes of Health, the largest granting agency in the US government when it comes to doling out the loot in science. It probably won't come as a surprise to you to learn that, on the whole, money is valued above most everythi1g else in research because it's funding that keeps labs like mine rolling alo1g just like money greases the wheels in other walks of life, so NIH is about as close to a higher power as you can get in science, the ones charged w th handing out the money and keeping the research establishment honest, deciding who gets what and who doesn't and on days like this one I often wish I'd gone into my second area of interest – history. But apart from the financial side of things, disease-causing bacteria like *Bacillus anthracis* are of interest to me because of this amazing ability they all have of enduring some of the harshest conditions that nature – and even some scientists – can throw at them.[ii]

And it's only within the last century that anthrax has been considered a rare disease. In fact, getting anthrax is one of the few things George Washington and Karl Marx had in common.[2] And one of my favorite writers, Ernest Hemingway, got anthrax when a few spores entered a cut on his foot while swimming in a river in France on his honeymoon.

Notoriety isn't anything new for this passive-looking bacterium as it had already secured its place in the "Biological Hall of Fame" long before 9/11 came along. Anthrax's rod-shaped cells were the very first microbes ever proven to cause a disease,[3] all by a country doctor in Germany (the MD kind) by the name of Robert Koch way back in the 1870's. There were no Petri dishes in Koch's lab in those days so he had to improvise by nurturing his anthrax cells deep within the sterile confines of an ox's eyeball. And as for an incubator, the good doctor made due with ordinary sand he heated with a kerosene lantern.[iii] [iv]

Whenever *anthracis's* natural surroundings become hostile and it starts to run low on vittles – animal flesh in this case – in order to survive it has this

---

[2] Anthrax was in the Americas before Columbus arrived. Washington got anthrax from his horse, while Dr. Marx (the Ph.D. kind) got it from some horsehair furniture. Well into the 1940's, horsehair was used as a binder for plaster in buildings, as workers discovered when they got anthrax in London's subway while remodeling the Kings Cross Station in 199E. More recently the Pentagon building burned for 3 days after 9/11/01 as firefighters tried to reach the fire. Water from their hoses was being blocked by horsehair used as insulation in the Pentagon. Jet fuel had penetrated all the way to the wood beneath. Some of the strongest memories of survivors are of burning horsehair.

[3] *Bacillus anthracis*. Technically, I should be calling it as an *endospore* since an endospore forms *inside* the cell rather than *around* the cell like a true spore does.

inner set of software, "Extreme Biology 2.0" I guess you could call it, a combination of genes that it can switch on (or off, depending on the gene) to transform itself from a run-of-the-mill rod-shaped bacterium into an even smaller, rounder, heat and drought-resistant spore all in the space of about a dozen hours, almost like going to seed, yet this is a single celled microbe we're talking about here.[v] This cell's ability to form these impervious little time capsules seems miraculous to me no matter how many times I see it unfold under the microscope or think about it while I'm drifting off to sleep at night, yet what often wakes me up in the middle of the night is the thought that this same talent for survival is also what bioterrorists find so interesting. For me, though, the payoff for working with these clever little bugs is the opportunity to understand life better at one of the more extreme ends of the spectrum.[vi]

Being small and rapidly dividing in two is a surprisingly simple strategy microbes hit upon over three billion years ago, which means that the ancestor of all living things discovered immortality long before Ponce de Leon ever thought of looking for the Fountain of Youth. About a billion years after the Earth's formation, the microbial world had already figured out a way to gain everlasting life simply by dividing in two endlessly...forever.

Bacteria like anthrax are survivalists that would put the human kind out here in Montana to shame and I'd be willing to bet there are few, if any, chemicals underneath your kitchen sink, including all the detergents and disinfectants, that could affect an anthrax spore in any observable way. While it's true boiling them in water can kill most of their spores, it takes about an hour to accomplish the task. You can torture them with strong acids or expose them to harsh ultraviolet light to your heart's content...all to no avail. They'll simply shuck you off as if you were a mild cold.[vii] But put these same intractable little beings inside a Petri dish with just the right amount of water and nutrients, stand back and watch, because their spores will come right back to life before your very eyes.

And I have no doubt that, out in the middle of some rancher's field not far from where I'm sitting right now, hidden within the topsoil of the sweet grass on the prairie, anthrax spores have lain dormant for 50 or even 100 seasons, just waiting for their chance, almost like Lazarus but they're not really dead as it's just *anthracis's* way of waiting around for conditions to change, its chance to steal back into the natural order of things after being taken up by the "walking around type of incubator," a grazing animal like a cow, of which my family owned several hundred head when I was growing up not far from here.

Anthrax has the talent to kill quickly, too, as researchers in Zimbabwe discovered when they came across a dead hippopotamus in the middle of the jungle. It seems anthrax had gotten there ahead and killed the poor critter so quickly it never even had the chance to hit the ground. They found the

6

hippo's carcass still up on all fours.[4]   And when two of the postal workers in Washington D.C. first began feeling sick during the letter attacks following 9/11, symptoms like a sore throat and difficulty breathing came on so rapidly that anthrax appeared even to a trained medical observer as if it were just a bad case of the flu.   One of the postal workers was even sent home from the emergency room with some Tylenol.   Needless to say, he was dead by nine the next morning.

It's because anthrax employs what we sometimes refer to as the "smash and grab" strategy.   Unlike some bacteria, anthrax doesn't rely on getting spread around by a semi-healthy, walking around patient who's coughing and sneezing all over the place.   Nope, anthrax plays hardball from the get go, dispatching its victim as quickly as it can, using this strong cocktail of toxins it brews up just for the occasion, then devours what's left of the body by making use of its enzymes, in the end forming hardy little spores again, then waiting down in the soil for a rainstorm or some wind to bring it back up to the surface so the whole ancient process can begin all over again.   And unlike most people these days, waiting is something *anthracis* does surprisingly well.[viii]

It helps to be tiny, too, even by microbial standards, when it comes to getting spread around.   You'd have to line up about 50 anthrax spores just to span the thickness of a single piece of paper; in fact they're so small they can penetrate right through the walls of an ordinary envelope, which is why it cost the EPA 30 million dollars to get rid of anthrax from the US Post Office's Brentwood sorting facility even though the total amount released was probably less than would fit on the tip of your little finger.[ix]

A good sign of a microbe's success in nature is when its lineage goes back a long way.   The ancient Greek doctor Hippocrates described what *anthracis* could do to grazing animals, and it's likely this bacterium is what caused the 6th and 7th plagues in Egypt's cattle as described by the Old Testament.[5]   And it was the philosopher Aristotle who recognized over 2000 years ago that anthrax came and went season after season in the same exact fields...but for reasons he could only guess at.   Not knowing anything about microbes or spores, Aristotle passed the whole thing off as being caused by rodents and lizards nipping at the cow's feet.

When I was growing up on my family's ranch – the Triple K – I used to listen to stories my grandfather would tell around the campfire about how back in the days of the great cattle drives down in Texas, back when the cowboys would bring their herds to the nearest railroad depot for shipment back east, how anthrax was so common back then that, even to this very day

---

[4] In a warm, blood & broth-filled Petri dish, *Bacillus anthracis* will divide to form two new cells about every two hours.
[5] Hippocrates is considered the first true *physician* because he didn't rely on supernatural forces to explain disease.

the spores that got carried along with the cattle back in the 1800's still reemerge from time to time...infections reappearing as if echoes from a distant past. Fortunately, we were far enough north that we never saw anthrax to that extent, and also fortunately for us, in order to make anthrax dangerous on a truly terrifying scale, their spores probably need to be treated with a powder so they can get dispersed into the air, or "weaponized" as we say, and that isn't easy either because being spread through the air is *not* part of anthrax's normal lifecycle.[x]

Even the most experienced infectious disease experts in the world, some of which I had the opportunity to work with back when I was a graduate student, are at a loss to explain all the peculiarities of *anthracis*, like why it is cats and dogs will get only a mild case of the disease, perhaps resembling a cold (if they get any symptoms at all), while these same spores can be so devastating if they find their way inside the gut of a cow or a horse, all of which are, of course, mammals. No one knows for sure even to this day.[6] In fact, one of the surprising things about bacteria is how boring they all tend to look under the microscope, even to a trained biologist like myself. Anthrax cells resemble little, rectangular boxcars all lined up bumper-to-bumper in a train yard but with nowhere to go and yet what these tiny microbes can do with their metabolisms is truly remarkable. A saying among microbiologists is that bacteria are interesting not so much because of what they look like, but because of what they can do. Some bacteria have metabolisms that will even allow them to dine on oil spills and monuments made out of the hardest stone.[xi]

But I don't want to waste all your time with this small talk, which is why I need to get back to the main problem that was on my mind that evening – how it was I came to lose out on the anthrax grant in the first place. I could shade the truth some by telling you that I was only interested in anthrax for patriotic reasons after 9/11, but the fact is, that was only part of it. You see, I saw anthrax as my best chance to gain tenure here at the university. And an assistant professor in a tenure-track position without achieving that brass ring after 4 years is...well you're pretty much out of a job is what it is.[7] Contracts like mine are only temporary, that is until we either gain our tenure, or we don't and have to go someplace else.

Unfortunately for me, the granting agency didn't seem to care about this small detail of mine. They seemed more concerned with the fact that my institution didn't have a BSL-3, or bio-safety level 3 lab, a highly secure area set aside for working with pathogens like anthrax or West Nile virus, microbes

---

[6] The largest number of people ever killed at one time by anthrax happened in Sverdlousk, Russia in 1979 when a worker removed a clogged air filter without telling anyone. As a result, 68 died when anthrax spores were released from the bioweapons facility the following day.

[7] In fact, I was officially classified as *contingent faculty*.

that can be spread through the air and so used as a weapon. Of course I realized the importance of containment and addressed the issue in my grant proposal as a footnote, which may have been a mistake looking back. The smallpox virus has been eradicated from nature since 1980 yet the last person to die from the disease was when a few of the virus particles got loose and floated out the window of a laboratory at a medical school in England, killing a photographer who was also working in the building.

BSL-3 labs are designed to contain germs like anthrax, preventing their escape with ingenious yet often surprisingly simple techniques like unidirectional airflow and workers having to walk through two sets of doors just to get in and out of the lab and things like that. I was aware of the containment problem and had planned to collaborate with a colleague at a nearby facility who happens to have a BSL-3 lab, but unfortunately the panel that reviewed my grant must have seen things differently. Maybe they thought 45 miles was too far away, I never asked. I have to laugh at the thought of 45 miles being a long ways away for government types back in Washington D.C. because even a couple hundred miles out here in Montana is considered "right next door" and if NIH had only known my high school football team used to travel twice that far on Friday nights just to play a single game, well maybe they would have reconsidered I'll never know.

But even two hundred years ago when the explorers Lewis & Clark passed this way they had backup plans for there was a bit of good news that came my way that day too. You see, the second grant I had applied for – my back up grant so to speak – was approved by NIH and due to this bit of good news I was able to keep my entire staff of two technicians and two students on, at least for the time being.[8]

Yet I still couldn't help thinking that maybe the real fly in the ointment that evening was that I wasn't all that fired-up about working with this second project, for this newer one involved the brewer's yeast and about the only thing I knew about brewer's yeast was that they could make beer and wine out of sugar and bread dough rise when you add water to flour and that's about it. Brewer's yeast aren't anthrax by any stretch of the imagination. But as my dear old grandmother still likes to remind my older brothers and me from time to time: "No person should ever turn down milk just because they can't get any cream."

I should also mention that I had picked up on a few interesting details about the brewer's yeast while researching the grant proposal, things I hadn't come across in any undergraduate biology class, like how alcoholic fermentation was the very first chemical reaction ever written down on a piece

---

[8] Lewis & Clark tried to have a backup plan in case the Native Americans turned hostile by camping on river islands whenever possible and anchoring their keelboat backwards to beat a hasty retreat upriver if & when the time came.

of paper back in the late 1700's by a French chemist named Lavoisier, or that the yeast is both a microbe and a fungus, meaning it's a lot more closely related to you and me than it is to any bacterial cell it might associate with down in the soil. In fact the unicellular yeast presents some of the same arrangement of molecules to the outside world that a bumblebee does on its surface. But in all honesty, I just wasn't all that interested. In fact, applying for the yeast grant was really just a way of keeping money coming into the lab if something should happen to the all-important anthrax grant, which, like I said, it did.

There were some other things on my mind too that day, like how missing out on all that anthrax money would affect my teaching load,[xii] and how it's possible I might even lose some of my bench space to another scientist who brought in more money than me this time around, but then I realized I'd probably manage. We Ph.D.'s tend to be a resourceful lot and what a lot of us won't tell you is that we all have our bread and butter projects in the lab, the tried and true ones we were brought up on because we learned them from our mentors back in graduate school bringing in money and publications almost guaranteed. [xiii]

And then on the side, in some out of the way corner, there may be a pet project or two on the back burner, the long shots we secretly hope will win us a Nobel Prize someday and these second projects are the ones we hardly ever talk about yet may have a student or two tinkering away on in the hopes of gaining enough preliminary data for a grant application in the future, but I need to get back to the brewer's yeast since the yeast is what precipitated my most recent adventure.

In spite of the fact that both are microbes, switching from a bacterium like *Bacillus anthracis* to a fungus like the yeast isn't all that simple. It's not true that all microbes are alike, in fact there's a surprisingly wide range of them in your mouth and under your feet and some that live in ponds and are so large they can even be seen with the naked eye. And if you could somehow gather up and weigh all of the living organisms here on Earth, the ones you can see with just your eyes, like trees and dogs, people and elephants, we would all actually get outweighed by everything you'd need a microscope to observe, microbes are that numerous here on Earth.

Well in nature the yeast – whose real name is *Saccharomyces cervisidae* by the way – is a loner, kind of like an amoeba or a paramecium, yet it's still a fungus, in fact it's more like a mushroom in the way it reproduces. The brewer's yeast even has a compartment like our own cells do called a nucleus to gather up its DNA inside of.[9] I like to think of the yeast as the "buzzards of the soil" because like other fungi they eek out a living waiting around for

---

[9] Bacteria are classified as *prokaryotes* because they don't have a nucleus.

things to die, things like berries falling off of a shrub or leaves off a tree, and then they consume them. But instead of claws & beaks, fungi come equipped with powerful enzymes, useful for digesting their food from the outside (unlike we humans who, of course, digest our food on the inside), and which is also why you might hear yeast and other fungi being referred to as "nature's recyclers" and while both are microbes and good at what they do, the similarities between bacteria and yeast don't extend far beyond that and no one's ever died of a brewer's yeast infection as far as I know.

Anthrax, on the other hand, is glamorous and lethal and scientists are people too so given the choice...well you're probably beginning to see my predicament already. Not only wasn't I very interested in the yeast, but my staff was even less interested, including the graduate student, and graduate students tend to get worked up about most anything involving science. And it's hard to keep the best and brightest at a university working for you when all you have to offer them is something that makes alcohol out of grape juice or bubbles rise in champagne when what they were really expecting was a BSL-3 level pathogen complete with plane tickets to seminars in far away places to give poster presentations about their work to really important people in the government.

So I spent the rest of what would turn out to be a rather eventful night wandering the halls of my department, alone down in the "dungeon" as we call our basement on account of how the only two places natural sunlight can penetrate is through two small windows at the stairwells on either end. I should also mention that we have these painfully long winters up here at this latitude and not being able to see the sun much doesn't help morale either.

Well I spent the better part of that evening feeling lower than a snake's belly in a wagon rut, trying to figure out the upside to all of this, and it seems that part of being a good scientist I've learned over the years is being a good salesman so I knew that if I could somehow interest my own group in the idea that working with the brewer's yeast is important in its own right, then I might just have a chance of keeping them on. But how could I if even I didn't believe it myself?

By midnight I was no closer to a solution and since everyone had already gone home for the evening, I was all alone. I should also mention that not many microbiologists wear cowboy boots so that night I made so much clomping around noises that the janitor came down to see what all the commotion was about and was visibly relieved to see it was just me.

So after reassuring Charlie and more out of boredom than anything, I found myself reaching for a packet of brewer's yeast – the kind you can find in any supermarket, the freeze-dried vacuum-packed variety[xiv] – and I added about a teaspoonful of water and a pinch of sugar, just like the directions tell you to do, to a few of the crumbs so I could have a look at them under the microscope, hoping for some kind of inspiration I suppose.   The name *Saccharomyces* literally means "sugar fungus" in Latin and in fact they do love their sugar – wherever they can find it.   Almost a third the weight of a freeze-dried yeast cell is going to be sugar, stored mostly in the form of glycogen. They keep this sugar in reserve for the chemical energy it contains, not so different than the way I keep gasoline in my pickup truck for its chemical energy.

I always like to start off with the clunky microscope, the heavy steel "boxy" one with the two wide eyeholes on the top so I don't have to do so much squinting when looking through it and, like I said, yeast are larger than a bacterial cell is so it doesn't take that much squinting just to catch a glimpse of them.[10]

It wasn't easy being alone in the lab that evening thinking about my missed opportunity with the anthrax project and I have to admit that few things are as uninteresting as a microbe that just sits there when you add water to it – rather ungrateful when you think about it – and just sit there is pretty much what the brewer's yeast cell does.   Lying in wait is part of its strategy, whether the yeast finds itself down in the soil or up on the skin of a ripening grape or the petals of a nectar-secreting flower attempting to bribe a passing insect into providing a free ride for its pollen onto the next flower. [xv]   Still, knowing all this didn't get me excited one iota.

And after a few minutes, every last one of the yeast had settled on their own accord down to the bottom of the droplet in front of me on that glass slide, which is kind of amazing in its own right because water to a microbe would be like us falling into a vat of maple syrup, water's that viscous to microscopic things.[xvi]   There's no worrying about brewer's yeast swimming off on you, though, no need to constantly refocus the microscope just to keep track of them because believe me...they aren't going anywhere.   You won't see any twitching or tumbling around, no gliding from place to place like an *E. coli* does inside your intestines as it whips itself around with its flagella to snag what's left of your breakfast, no crops of cilia along its outside like a *Strombidium* has to twirl itself around until it gets to where it's got to go.[11] Nope, they won't even ooze about like an amoeba does on a rock.   Instead these little round specks just sat there imbibing sugar, as if mocking me, as immobile on the bottom of that glass slide as my own career had suddenly become on land.[xvii]

The very first person to ever see bacteria lived in Holland and his name was Antony van Leeuwenhoek.   Back in the 17th century Leeuwenhoek was

---

[10] Indeed, the average microscope throughout the late 1800's was stronger than the one I was using that night – but mine is more comfortable.

[11] Some *Vibrio* bacteria can travel 100-times their own body length in under a second using a rotating flagella for a propeller (it would be like us trying to run 300 miles an hour on land).

also the first to catch sight of a yeast cell, too, and it turns out that even the "Janitor from Delft" wasn't all that impressed. In fact, he didn't even think yeast were alive. Like me, Leeuwenhoek was partial to his bacteria, once referring to those he examined from his mouth as "little beasties" because they were so swift while twirling about "like a top" and "prettily a-moving" as he put it. Yep, that Dutchman sure had a way with words.

And even though it's relatively gangly by microbial standards (50 yeast cells lined up end-to-end would just to span the diameter of the dot above the letter i) I kept thinking to myself how far I was from where I'd expected to be when I first left graduate school just a few short years ago. Here I had been trained to work with bacteria, and not just any bacteria but the kind that cause devastating diseases. And pathogens are infinitely more interesting than a yeast cell is to someone with my background. Yeast don't secrete toxins of any sort, certainly not the kind some bacteria can use to crack open a red blood cell to steal its iron or make someone cough up phlegm from their lungs. Nope, being a non-pathogen like the yeast means having none of these strategies at your disposal. Well the fact was, I just couldn't get excited about the yeast in the least and I had to...soon...because fall semester was getting ready to start and I was in danger of losing my entire staff.

THE E-TICKET RIDE

It's a funny thing how eyestrain can set in a lot sooner after you've had a rough day, and so after a few minutes of squinting I gave up on the yeast and logged onto the Internet in the back room of the lab, the small place we have set aside for two computers, a large incubator, and a sterile hood for streaking microbes onto the surfaces of agar dishes. So I clicked on it more out of boredom than anything, first to my e-mail, then I found myself at that giant encyclopedia in the sky, the one with just about everything in it if you care to look, and with two fingers typed in the words "brewer's yeast".

All I wanted was to put my mind at ease so I could get some sleep. Well at first nothing happened, so I followed a couple links that lead nowhere in particular, then gave up and did an image search. It was then that I came across something to hold my attention and it was unmistakable. I had found the most amazing photograph of the pyramids...all three of them, too...looming high above the desert floor with an orange, puffy sun setting in the background. When I scrolled over it, the caption read simply, "Brewer's yeast floating in the Nile gave the world its first leavened breads around 2600 BC."

Well like I said, history was going to be my second choice of study, so I lingered there for a spell just to see if I could figure out what the connection was until it dawned on me I had stumbled onto a website devoted to the history of civilization. Still, I couldn't help wondering: what did the dawn of civilization have to do with a fungus like the yeast? I got about a paragraph into it when this plane trailing a banner flew right over the top of the pyramids and so I clicked on it.

Now I consider myself a person of science even though I still hold out hope of a higher power having a hand in things here on Earth, and as it turns out this plane belonged to some travel agency way off in New York City. And my vacation was fast approaching yet I was in no mood to spend it like I had the previous three summers – working down in the dungeon – all the more reason to hit the back button on the browser and resume reading, this time looking for an excuse to go someplace.

The website explained how the ancient Egyptians were the first to turn the leavening of bread dough into an industry. In fact, Moses and his followers when they left Egypt missed their leavened bread so much that one of the first things they did after wandering around in the desert for so long was to start growing wheat so they could bake bread again just like they'd learned to back in Egypt. Apparently this process had been perfected by the Egyptians, who then went on to invent some 40 different kinds of bread.

So the Egyptian knack for raising dough seems to have rubbed off on people from time to time. I'm always teaching students new techniques in the lab myself, whether it's how to purify bacterial toxins using an affinity chromatography column or the proper way to store glass beakers on a shelf without collecting any dust. This was all interesting and yet the failed anthrax grant kept drifting in and out of my mind that evening. The travel agency had mentioned something about a limited time offer, so I clicked on the plane again as it buzzed around a second time, this time a little more wide-awake and determined to flesh out the details.

For most folks there comes a time during their formative years when they run across a book they can still remember. For me it was *The Innocents Abroad* by Mark Twain.[12] I'll just mention how this thin little book is basically a compilation of stories written by Twain back when he was traveling through Europe and the Holy Land and that, even though it's a century and a half old, still manages to transcend time. I had missed out on my chance to go to Europe and see all the great ruins while everyone else my age was going mostly because I was too busy volunteering in a lab to get some experience I could put on an application for graduate school (science is competitive). The farthest I'd ever ventured was up to Canada, and that was only one time to cheer on one of my brothers in a rodeo. Come to think of it sitting in my chair

---

[12] I would learn later that Twain was no stranger to bacterial diseases, having lost two children to them, one to meningitis, the other diphtheria.

that night, I realized I had never gone off anywhere on my own that was very far away in my whole life except to graduate school.

Mark Twain arrived in Egypt back in 1867 when he was about the same age I was now – on a Civil War era steamship called the *Quaker City* – among the very first to take part in such a journey back to the Old World – the very first American tour group you could say – and Twain wrote about it for a newspaper back in San Francisco.[13]

I scanned the fine print on the screen, hungry for more details. The ad claimed the ticket was just $550. A measly few hundred bucks for a round-trip flight all the way to Egypt and back? Just as sure as a goose goes barefoot, there had to be a catch. I could feel my heart make an unscheduled appearance in my throat as considered the possibilities. It even promised tax was included.

I thought about the travel guides scattered among the textbooks gathering dust on my shelf across the hall in my tiny office. One was a guidebook for Nepal, a place I never expected to see. I just liked to take it down and look at the glossy photos once in a while. I rubbed my eyes again, unable to blame my worsening eyestrain on the yeast, for they had only managed to resurrect what I'd already gotten from grading so many final exams the previous week. It turns out it's practically impossible to design a biochemistry exam in multiple-choice format, which is why they always end up looking more like essay tests.

This was getting sillier by the minute, I thought. What was a middle-aged professor whose chances for tenure had just been diminished by a couple orders of magnitude going to do wandering around Egypt? But human nature being what it is I just ignored all of this and kept on going, entering some dates off the top of my head, fully expecting the airfare not to be there anymore. Then I clicked on the "Begin Search" button and within a few seconds the screen changed to black, everything except for that little hourglass thing, and so I waited...and waited...and then waited some more until finally a computer-generated image of a plane ticket materialized in front of me. It even had its own bar code, and most amazingly of all, the fare was still $550. All that was left for me to finalize the agreement was to enter my name and a valid credit card number along with that 3-digit code on the back.

My fork in the road that night had materialized in front of me in the form of an airplane with a cartoon face, one that was now standing on its tail, winking, and using its wings like hands to prompt me for more details. Window seat or aisle? The most I'd ever bought online before was a country & western CD

---

[13] The Alta California, which also sponsored his trip. Specifically tailored tours of Europe were not new, however. Thomas Jefferson, for example, took an English garden tour in 1786, which would have an influence on the design of his own gardens back at Monticello in Virginia.

so I chose the window seat, took a deep breath, wiped the accumulating sweat on my palms off onto my jeans, and then clicked the accept window seat icon. The screen went blank again, as if commiserating with some higher power for advice on the matter, and then after what seemed an eternity the plane flew back across the screen trailing a banner that read, "Congratulations, Dr. Ketchum. You have just purchased a round-trip nonrefundable ticket to Cairo, Egypt. Please print a copy of this for your records. Would you like a rental car when you arrive?"

I wish I could say I had decided at that moment to go, but the truth is, I didn't know until somewhere between school and my apartment on the long walk home that evening when it occurred to me that in less than a week's time a jet bound for Salt Lake City would be taking off from Missoula, Montana with or without me on board, then a second seat assigned to yours truly and already paid for would arrive in Frankfurt, Germany and finally a third would venture all the way to Cairo, with someone else sitting in it if I backed out, that I decided to go. And I decided it all without having the foggiest idea of just how it would help me with the new project. The final nudge came with the realization that a two-week trip might be just the thing to get me caught up on some pleasure reading, but without the eyestrain that always accompanies grading exams.

While shuffling through the leaves alongside the road that evening I had plenty of time to think about just where it was I found myself at this point in my life, which wasn't good by hardly anyone's measure. Feeling the cool dampness of the fall leaves as they worked their way up above the tops of my boots, my mind began wandering, recalling how fascinating Egypt had always been to a high school student in Thistle, perhaps because it lasted so much longer than any other civilization.

The ranch I grew up on was less than an hour's drive from campus, yet it may as well have been a world away these past couple months. Somehow I had slipped into a place I never knew existed, an unfamiliar place I hadn't grown up in or knew hardly anything about, a place I felt more and more an outsider in every day. Not many ranchers' sons spend all their free time thinking about tenure. At the very least Egypt offered a chance to find a new perspective, like standing on higher ground to survey the surrounding terrain better maybe. Egypt was the ancient world's granary. Thistle, the town I grew up near and home to my university, is the granary for the entire Lewis Valley. But unlike here where the valley is walled off by mountains on all four sides, Egypt was in a river valley protected by deserts. And if it wasn't sand

and lack of drinking water blocking invaders back then, it was the Mediterranean Sea to the north.[14]

Yet when Jesus walked the earth some 2000 years ago, Egypt was already ancient. The great pyramids had already been standing for 20 centuries before the Roman Empire began running things, and to Greek travelers like Antipatros, the mystery of how the pyramids were built had already been lost to time.

After opening the door to my small apartment I was greeted by about a feedbag's worth of affection launched into my midsection, his two front legs draped over my shoulders, the momentum of his body sandwiching me up against the wall while a sandpaper tongue was scouring my face. Barney is a white boarder collie I'd raised as a pup back on the ranch. The black patch over his eye made him look like a pirate, and he was definitely getting older, which is why I sometimes called him a senior pirate, and he had less energy than he used to and so was more likely to be found taking long naps by the furnace grate on the floor in the kitchen when I arrived home than looking out the window for me, but mostly I never noticed. Dogs, not horses, were the very first animals ever domesticated by humans and I've never doubted as to why. They've been by our side since before we took to agriculture even, there to protect us. After walking Barney once around the block we both welcomed sleep.

BIOPROSPECTS

The next morning, after discussing it with some of my colleagues, a consensus had emerged down in the dungeon that the reason I had gotten such a good deal the previous night was that I was willing to fly last minute. Apparently I had stumbled onto the fact that airlines would rather sell tickets at greatly reduced prices online anonymously, than to publish lower prices, which could trigger a fare war among the airlines.

The next several days went by slower than a hay wagon hitched to a hog as the whole thing had taken on an air of incredulity to it, as if it were happening to somebody else. Of course my family was surprised and it helped bring the reality home some when I phoned them with the details. Even so, I still had to take the ticket out once in a while just to see my name on it. Thoughts of the anthrax grant were fading, replaced by things I hadn't

---

[14] Egypt's relative isolation also explains why it didn't need a standing army to defend itself during the Old Kingdom period, which lasted about 500 years, and why hieroglyphics would someday die out as a writing system. For nearly 2,000 years not a single person alive could decipher them, not until after the Rosetta stone was discovered by Napoleon's traveling army of scientists in the late 18th century.

considered since I was a kid, back before I knew what a grant application looked like, and since it's possible to get a visa at the airport in Cairo upon arrival, I had only to figure out a way to keep my lab up and running, as well as decide between the two backpacks in the hardware store window, either the one with the strap across the waist for added security, or the burlap one with no strap but was at least lighter in color and so better equipped to reflect the Saharan sun. In the end, I decided on the more secure pack, which probably says more about me than anything.

My last day before leaving the butterflies were still coming and going in spite of the fact that, twice a week another professor had promised to check on things like the temperature of our incubators and the levels of liquid nitrogen in the cryogenic tissue culture jugs, requisitions for the new yeast project would be placed in their appropriate wire baskets in the office upstairs by my two technicians, plus I knew I had some of the best in the department working for me, so what could go wrong? As someone wise once said, you know you have a good horse underneath you when you can forget all about the saddle. Or maybe it was that a good horse is seldom spurred. I'm not sure which applies best here.

As lunchtime rolled around I found myself brushing up on some ancient history, ham sandwich in one hand, guidebook in the other, eager to take down some notes, looking for any clue that might get me started with the project. I had forgotten how the pyramids were built by some 20,000 farmers fed a steady diet of bread and beer delivered to them on site, in fact the pharaoh even had an entire city raised near the pyramid just for providing his workers with these staples. I'd also forgotten how wine containers were found alongside King Tut's mummy to quench his thirst after death, complete with information describing the wine's vintage and place of origin.[15] How strange to think that the very same microorganism that once nourished a nation was what my own future could very well depend on now.[xviii]

I laid the book down on my lap and considered how little time I had left now. I was no kid anymore; in fact I wasn't even the youngest assistant professor in the department. Thomas Jefferson authored the Declaration of Independence when he was just 33 and Napoleon was all of 29 when he led an entire army off to invade Egypt. And the Frenchman who would someday decipher the Rosetta stone, Champollion, already knew Greek and Latin and was busy studying Coptic, Hebrew, and Arabic...all by the age of 12. Tenure committees like mine are made up of other scientists and in a year's time these eight would either recommend to the provost that I be made an associate professor and given all the rights and responsibilities associated

---

[15] Some of the bread in King Tut's tomb still had fruit in it...tangible evidence that pharaohs craved variety in the afterlife. / Ancient Egyptian doctors sometimes used moldy bread as a poultice to prevent wound infection.

with that honor, including my tenure, or that I should be let go. There was, in all practicality, no in-between.[16]

My last afternoon before leaving was spent moseying around the lab, trying to put things back in order, looking for anything I might have missed, wondering if Twain had felt this nervous before he left for Egypt, but then I recalled how Twain had already been to Hawaii and back by then (folks called them the Sandwich Islands in those days).

A quick glance through any microbiologist's freezer will turn up all sorts of curious artifacts: projects that had once been at the forefront of one's thoughts but were now relegated to thimble-sized plastic cryo-vials, the kind with screw-top lids and rubber gaskets for keeping out the air. Or in the refrigerator there might be forgotten about microbes growing on Petri dishes, their once-separate colonies now having merged, becoming one continuous lawn of microbes, no longer content to live as separate little spots; other dishes having dried out in spite of the wax paper stretched around their outsides in a vain attempt to keep the air out and the jelly from losing its water and then shrinking, forming a network of little cracks, like those you see at the bottom of a dried up mud puddle. Another shelf was filled with glass test tubes waiting to be inoculated...or contaminated...whichever came first, perhaps by some fungal spore from a distant mushroom taking advantage of the day's local air currents.

We microbiologists are all collectors at heart, not so different than the way an entomologist collects bugs and then displays them on pins, or the way a coin collector is always on the lookout for rare specimens in his change at the supermarket, we microbiologists are never that far away from our next find either.

It might be something from right under our feet, about a hand's depth in the soil, or floating around in the air above our heads, or even lingering on the outside of our bodies, or inside the gut of a cow; maybe even on the surface of a fern. Microbes are everywhere you turn in nature and insulated kettles of liquid nitrogen are the mantles above the fireplace we display our trophies in; perhaps a species of bacteria we worked with several years ago on a project that never went anywhere, the microbes themselves now lying in suspended animation, as if waiting for the next idea to hit us before being thawed back to life again.

As for me, having cut my teeth bio-prospecting in the thermal features of Yellowstone National Park, it's only natural that my first inclination was to learn all I could about hyperthermophiles, the kind of single-celled life forms with names like *Aquifex* and *Thermotoga*, inhabiting waters so hot you

---

[16] Sacajawea was just 16 when she helped guide the Lewis & Clark Expedition to Montana. Sitting Bull – Custer's rival at the Little Big Horn – would count his first coup in battle at 14. Custer was promoted to brigadier general at age 23 during the Civil War.

wouldn't dare stick your hands into,[17] bacteria once attached to pebbles on the bottom of a hot spring, now stored silently away in a plastic vial and labeled with a thin black sharpie complete with, not only the name and date they were found, but also the substratum they were collected on, in geologic formations like "Whirligig" or "Steamboat Geyser" or "Grand Prismatic Spring". This part of the American West is different from the rest of North America for a lot of reasons, not least of which is that the crust is thinner here and so the hot liquid magma from down below has a chance to rise up closer to the surface than it normally would, nearer to where the rainwater and snowmelt from the mountains can get at it.   This water then becomes super-heated underground, the Earth's thermal energy making it expand and begin its rise back up to the surface.   It's all part of Mother Nature's perfect, time-honored recipe for a geyser.

On another shelf in a plain white freezer was a bacterium I'd come across while working in an abandoned sub-surface mine near Butte, Montana, as inhospitable a place if there ever was one for a living creature, with water pools 1000 times more acidic than what's inside a car's battery, strong enough to dissolve metal tools left inside them overnight by some overworked graduate student.[18]   Yet the microbes I was holding in my gloved hand would have preferred this harsh acidity to the balmiest beach in Hawaii. Extremophiles are a reminder of how even the scientific literature can be biased from time to time because to a microbe like this one, it is *we humans* who inhabit the extreme environments.   The oxygen we take into our lungs whenever we inhale and that bathes our skin continuously is a toxin to so many different microscopic bacteria, most of which we'll never know about.

A feeling of regret washed over me upon recalling that bioprospecting would be out of the question on this trip.   Things had gotten too difficult after 9/11.   Prior to this it wasn't unusual for microbiologists to carry our finds around, distribute them at meetings in ordinary envelopes stashed away in our top pockets, freeze-dried and harmless like so many grains of coffee lying there lifelessly.   I once sat through an entire talk by the biologist Lynn Margolis – in fact I can even remember her topic, which was on the evolution of mitochondria from bacteria – with an envelope filled with bacteria collected from rustsicles on the hull of the Titanic.   This particular microbe had kept its appetite for iron over millions of years...and the person I'd gotten them from wanted me to help his company find better ways to remove dissolved iron from drinking water, so he had the microbes lyophilized (freeze-dried) into a

---

[17] Near-boiling at over 80° C (176° F).
[18] Most bacteria are killed by acid and some historians have speculated that the reason Lewis & Clark's men suffered from dysentery for much of their journey was that they didn't add vinegar, a weak acid, to their drinking water even though it was a common practice in the early 1800's.   Why they didn't is still a mystery.

coarse dry powder which was by then resting harmlessly in my top pocket during the talk. No one was ever the wiser.[xix]

It's not hard to revive microbes usually. Once back in the lab just a small amount of sterile water and the right choice of carbon for an energy source, usually glucose, some ammonia salts so it can manufacture amino acids, that's about all it can take, that and a modicum of patience. Microbes make a living as opportunists,[xx] not so different than scientists do these days come to think of it. We have to be with funding the way it is. Today, the red tape alone would be enough to stop all but the most foolhardy from bringing back a microbe in their top pocket from overseas, and with my tenure on the line I sure as hell wasn't going to be accused of carrying anything suspicious back from the Middle East. I had enough to worry about as it was.

It just dawned on me that I never did explain how I came by the yeast project in the first place, the reason I was awarded the grant and was now on my way to Egypt. Why would NIH give an assistant professor like me – and a tenureless one at that – $250,000 to study yeast if it wasn't to make a better jug of wine or loaf of bread?

Well the answer is simple enough. The brewer's yeast – like the bacterium *E. coli* that lives inside our gut – has become one of the workhorses in modern microbiology labs over the years, useful for making all sorts of human proteins inside of.[19] They're miniature fermentation vessels, each yeast cell its own test tube, delineated not by glass but by a fatty cell membrane, tiny cells that double as fermentation factories for the assembly of valuable medicinal proteins. Microbes like the yeast can be coaxed into making all sorts of human proteins like insulin, interferon, and antibodies for example, which is a lot cheaper and easier than trying to acquire these same proteins from human tissue. A whole heck of a lot easier, believe me.[20][xxi]

The downside is that, because these proteins have to be purified away from the bacteria or yeast, often times pesky chemicals can slough off these microbes and co-purify along with the human proteins you're really after. And if these stray contaminants get injected into a patient then all hell can break loose. They will get recognized by special cells we all have whose job it is to constantly go around looking for things that don't belong in the body and then set off inflammatory reactions when they find them, complications that can be serious, even deadly on occasion.[xxii]

My grant involved making a trial human protein of my choice, then genetically modifying the yeast so it wouldn't produce so many of these stray

---

[19] I knew yeast were being used to make non-yeast proteins because of a paper I'd come across by some structural biologists growing anthrax toxin (which normally functions as a set of 3 different bacterial proteins) inside yeast cells.

[20] Most human insulin used by diabetics today is made inside genetically engineered *E. coli*

yeast molecules in the first place. Pharmaceutical companies spend millions to purify human proteins away from bacteria and yeast to guarantee they are free of contamination and if we could genetically engineer a brewer's yeast cell that made even less of its own molecules to begin with, well this would be of benefit. A tall order, *humanizing* the yeast, but the amazing thing about the brewer's yeast, as I would soon come to discover, is that it has been so amenable to change over the years.

It's one of the few microbes in the lab where you can actually add a stretch of human DNA, say the gene for insulin, and the microbe will take up this gene and incorporate it into its own chromosome. Chemically, DNA is DNA throughout all the kingdom of life, whether it's DNA from a bacteria that causes anthrax or an amoeba, a bird or a dinosaur. It's the *sequence of the bases* that makes DNA different between different organisms, and in fact microbes will make the corresponding protein the gene codes for just as if it were making the protein for its own self. And that's about all you have to do to get your foreign DNA inside a yeast cell, just add it to the yeast culture along with a quick jolt of electricity, which is pretty amazing when you think about it.

But this kind of research is more along the lines of "filling-in-the-blanks" than actually discovering something new and completely different and therefore it's never as exciting. A lot of research is like this, though. Few of us have ever won a Nobel Prize just by filling-in-the-blanks in science. Nobel Prizes are usually given out to those not afraid of going out on a limb once in a while. And so I was following an inner voice to Egypt that, deep down told me if I could just learn something new and interesting about the brewer's yeast, uncover something special, then it might give us a leg up with the new project and light a fire underneath all of us to accomplish something important.

A RIVER IN EGYPT

"Egypt," as the wise man said, "is the gift of the Nile." Or maybe it was "Egypt is the Nile," I couldn't recall exactly as the cabin crew began its preparations for landing. Merely crossing the Atlantic had been the longest trip of my life so naturally I was feeling the excitement that always builds whenever you begin your decent someplace new. The ancient Greeks were first to call where the Nile empties into the Mediterranean a delta because their letter "delta" was shaped like a triangle except that the triangle below me probably delineated the first place humanity ever used irrigation for crops.

I watched the flight attendants collect all the leftover drink containers as our plane droned on towards the place where perhaps the oldest civilization in human history began, having begun my journey in one of the newest.   Up until the Eiffel Tower, Egypt's Great Pyramid was the tallest structure ever built by humans, a feat that was to remain unrivaled for four thousand years. In fact, the pyramids are all that's still standing of the original Seven Wonders of the Ancient World.[21]

Without warning, the plane veered sharply to the right and then dipped as I was beginning my way towards the rear of the plane to take one last look in the mirror.   I continued risking my neck because I was to be met by a woman doctor named Tahany Hassan (the MD kind) from the American University in Cairo.   And as I'd already seen a photo of her on the Internet I expected she would be attractive.   While trying to find my seat again, a new wave of anxiety surged through me, not because of Tahany but for my own safety, as the plane dipped yet again, then a third time, finally leveling out just low enough to make out the long columns of lights marking Cairo airport's runway.

At least I had no baggage to claim, I reassured myself, having stuffed everything I could think to bring inside my new backpack.   I'm still not sure why I thought the markets in Cairo wouldn't carry toothpaste.[xxiii] With my feet touching the pack, I leaned further back into my seat and reminded myself that I too was a doctor, even if it wasn't the MD kind. That should count for something in her eyes.

To get a visa involved a simple cash transaction and a cursory glance at my virgin passport; my first welcome sign coming in the form of an official stamp.   Feeling a strange sense of kinship towards my fellow passengers, I followed those in front of me through a turnstile, backpack slung over my shoulder, turned sideways to make it through the rotating stall and was suddenly greeted by my first real surprise in Africa...a gust of sticky, warm air. Even deserts can be humid.   My excitement continued as I pushed through a gauntlet of turbans and beards – foreign faces holding up handwritten signs advertising hotels and tours, some with passenger's names scrawled on them.

I wiped the sweat from my forehead and used it to dampen down the place where my hair always has a way of standing straight up from in every photograph I've ever taken.   Tahany was probably in the crowd somewhere and she didn't know what I looked like yet.   I deliberately hadn't gotten around to putting a photograph of myself up on my own department's website. Like I said, I've never taken a picture I've ever liked.

Cairo, like so many ancient cities, has accreted over the centuries, reminding me of the way a snowball does while rolling down a hillside. Newer portions were added on as more recent arrivals saw fit, and when the Moslems began their turn running Egypt in AD 641, they founded the walled

---

[21] There was a brief period of only 60 years when all Seven Wonders of the Ancient World were standing at the same time.

city of Al-Fustat, what is today *Islamic Cairo.* A little further to the south, Egypt's Coptic Christians – here since the days of St. Mark – had already built their own version of a city with some buildings surviving into the present, as does the overall layout. *Coptic Cairo* today picks up where the ancient remains of an even earlier fortress built by the Romans left off, and I had plans to visit it the next day if my jet lag allowed. But even in the Middle East things change, albeit more slowly compared to a lot of places, and by the mid-1800's an even newer version of Cairo had emerged, one like what you'd expect to see in Europe today, in fact it's why Cairo is sometimes called the "Paris on the Nile". The solution turned out to be straightforward enough. To make room for the new addition, Ismail the Magnificent simply had some unused swampland near the Nile northwest of the old city drained away, and because of his decision this is where *New Cairo* and my hotel sit today.

With a population of just over twenty million (coupled with the fact that it was never conquered by the Mongols), Cairo is easily the capital of the Arab world. In fact, few countries are so dominated by a single city as Egypt is by its own capital. And as I was discovering on the way to my hotel that evening, one in every two cars in Egypt is also taking up space on one of Cairo's congested streets somewhere.

I gazed at the exotic script on the billboards advertising cigarettes and Coca Cola, still looking for more tangible evidence of being someplace far away during the drive south from the airport when my driver – a young university student named Mustafa – pointed out the Arab League Building to my left. I should probably have mentioned that it was one of Tahany's medical students, Mustafa, and not Tahany herself, who was the one holding up the sign with my name on it, and Mustafa who felt the need to apologize for his boss's absence, while I looked down at the ground in an effort to hide any disappointment that might have shown. He had been given the task of explaining why his mentor had been called away at the last minute. One of her patients was in the midst of having a baby, yet I still couldn't help wondering if maybe she hadn't found an old photo of me while doing a Google search. I went through a period where I was a lot heavier than I am now back in graduate school.

During the drive Mustafa explained that, further out in either direction from the Nile, are situated Cairo's suburbs with names like Giza and Heliopolis and Shubra, the last one being where Tahany and her extended family lived. Mustafa, it turns out, was unusual in that he lived alone in a small apartment in Giza, a suburb that stretches almost the entire ten miles from the outskirts of Cairo all the way down to the Great Pyramids to the south. With fully ninety percent of her population living within just a few miles of her banks on either side, Egypt is still very much a gift of the Nile.

As we neared the pulsating heart of New Cairo we came upon Tahrir Square, which looked more like a circle from my vantage point, and the smell of exhaust became palatable through the open window. My pulse quickened as the carbon monoxide made its presence felt in my brain. I was getting a taste of what it must be like to be thrust into a time machine of sorts, of being very far away from one's comfort zone, of what an acrobat must feel when he looks down and suddenly recalls that he's been working without a net the whole time. Adrenaline produced above my kidneys entered my bloodstream, teaming up with the carbon monoxide already coursing through there as we rounded the bend that thrust us into a giant beehive of activity. We found ourselves in the midst of what can only be described as frozen chaos.

All around was a life-sized panorama of faded busses, passengers hanging out of open doorways, motorbikes with extended families squeezed onto gas tanks, a specter of smoke hanging over everything and reflecting the headlights, in the background some Arabic tunes emanating from some place indeterminable. I'm still not sure how many lanes there were because there were too many vehicles converging from all directions to see any lines, motorbikes impatient with the pace of traffic squeezing between busses and dodging pedestrians taking their chances, the exhaust causing my eyes to water over, me looking out the window and wondering how anyone could go through this ritual every day. A feeling of vulnerability sent me reaching for my backpack to make sure it was still down beside my foot, as if it could have walked off on its own. The fact that Mustafa didn't seem fazed made the scene seem all the more surreal, but after a good ten minutes the gridlock broke apart like so many ice sheets do on the Yellowstone River in springtime and Mustafa located an opening and rushed down a side street leading to my hotel.

The Bedouin had come highly recommended by Tahany in one of our string of e-mails, perhaps because her brother ran it, or owned it, I still wasn't sure. When she'd mentioned how it had once been used by Egyptologists back in the 19th century, I made the reservation and besides, it was supposed to be a stone's throw from one of Cairo's most important thoroughfares, the Sharia Talaat Harb,[22] a section that experienced a building boom back when the Suez Canal first opened in 1869, two years after Twain and his party rode the train here from Alexandria. I was promised "a single shared balcony overlooking a bustling avenue." The room had this, as promised, but otherwise turned out to be rather plain, accompanied by a single wooden table and chairs, a musty odor, a double bed and that was about all I can recall now. The only other concessions to the world I left behind were a

---

[22] My guidebook described Cairo's older streets as "living museums".

ceiling fan, a telephone that never seemed to work when I needed it and a western-style bathroom I wasn't required to share. Egypt, I was learning, has a noticeable lack of hotels in the mid-price range, as the tourist industry has grown up for so many decades around cost-conscious backpackers.[xxiv] These days at the other extreme are the newer five-star extravaganzas with names like Nile Hilton and Sheraton Cairo. Still, having traveled halfway around the world I wasn't going to allow a heated swimming pool and exercise room into my recollections when I'm old and gray.[xxv]

After checking in I laid on the bed face-up, the balcony door wide open, staring up at the ceiling fan that was making inroads at replacing the musty smell, soothing my exhaust-stung eyes as they followed the fan blades twirling fast enough to remain a blur, barely able to convince myself I had actually made it…and all in one day. Twain wasn't the first tourist here and of course I wouldn't be the last. Egypt would come to remind me of one of those flowers that manages to emerge from the crack of a busy city sidewalk and bloom against all odds. The entire civilization sprouted out of the world's largest desert – above a narrow patch of land fed and watered annually by the Nile – yet Egypt was also a tourist destination as far back as the ancient Greeks and Romans. In fact, it was the Greeks who first paid homage to Egypt by referring to her as their own "cradle of civilization", a complement coming from a people that considered all other outsiders barbarians. In much the same way someone from my own world might take a trip to Europe after finishing college, the future movers and shakers of ancient Athens and Halicarnassus – young and wealthy men like Plato and Herodotus – paid a visit here to take in some of Egypt's wisdom and bring it back to enrich their own corner of the Mediterranean.[23]

Egypt was where the ancient arts of medicine, astronomy, and mathematics had already been incorporated into daily life long before a Greek traveler named Pythagoras "discovered" the formula for describing a right triangle here. More recently Napoleon's three-year sojourn into Egypt beginning in 1798 – part military and part adventure – would soon reveal Egypt to an astonished Europe. Napoleon took along 167 artists, scientists, and scholars to record not only the major finds but the details along the way, all of which, the terrain and ecology included, were dutifully depicted.[24] However, the spike in tourism Egypt has most recently experienced can be traced to a single event that took place on November 16, 1922, which is when

---

[23] Herodotus, the world's first historian/tourist, wrote about Egypt and as a result, Greeks that came to Egypt after him (during Aristotle's time, for instance) wanted to see the same sites Herodotus had described. / Many ancient Greeks were convinced their own civilization had its beginnings in Egypt.

[24] He was in Egypt hoping to cut Britain's trade route to India, and besides, being Napoleon he probably enjoyed the idea of traveling in the footsteps of Alexander the Great two thousand years earlier.

King Tut's tomb was discovered by an artist with no formal education in archeology, a painter by the name of Howard Carter, and he and the boy king would generate headlines around the world for months. [xxvi]

I laid the guidebook back down on the bed beside me and looked at my watch. It was almost midnight and only now were the first signs of fatigue invading my frame. I took another sip of my tea...then cradled the cup back in its saucer on the bed, imagining what it must have been like for Twair's party in the 1800's. The first two weeks onboard their floating hotel – a side-wheel steamship – had been spent just getting from New York City to the Azores west of Portugal, where they all had to stop and recuperate from seasickness. Their next destination was Gibraltar and eventually Egypt. The *Quaker City* while crossing the Atlantic had taken an enormous pounding from storms I could barely imagine while I, on the other hand, had made the entire journey in just under thirteen hours and two plane changes, and was only now succumbing to indigestion from my umpteenth bag of smoked almonds taken on an empty stomach. In the mid-1800's there were several guidebooks Twain and his party could have chosen from, most offering more in the way of romance than practical advice, always somber in tone a few even instructed the reader on how he or she was supposed to feel while staring up at the great ruins of antiquity.[25] To his credit, Twain made up his mind early on to use his own experiences, one of the first travel writers to interject personal humor this way into a travel guide. I couldn't help thinking that, in spite of the gulf in time between us, that I had come here for many of the same reasons Twain and his companions had...adventure...and a chance to look at my everyday world in a new way after returning home.

Just as I was drifting off, the sense of helplessness that sometimes accompanies sleep startled me back awake. It was the crowdedness of it all.[26] Cairo at one time had been a wealthy city situated at an important point along the Silk Road, home to a wide assortment of traders dealing in rich spices and perfumes and other novelties from the Far East.[27] It was no wonder perfumes were important back then, I thought. What with all the heat and humidity things had to have gotten rather smelly, still it was strange but in all that 21st century gridlock Mustafa and I had come upon on our way to the hotel, I didn't notice a single face that looked angry or upset in any way.

---

[25] According to historian Stephen Ambrose in his book *Undaunted Courage*, the journals Lewis & Clark kept of their trip across the continent can be considered the first guidebooks written west of the Mississippi.

[26] Only 4% of Egypt's land is inhabitable (about the size of Maryland), and yet this is where 80 million live today.

[27] It was the rise of the Ottoman Empire and their monopoly on the spice trade that led the Italian explorer Columbus to set sail for India in 1492.

AS WIDE AS THE MISSISSIPPI

I only realized I had left my balcony door wide open when my sleep was being replaced by the Moslem call to prayer. I slept through the traffic noises as the sun was busy making its appearance, nudging itself above the red tiled rooftops, the call to prayer enveloping the streets and narrow alleyways below me, filling what looked from my vantage point like numerous capillaries branching off its main artery feeding the Sharia Talaat Harb with an endless stream of cars, busses, and motorbikes. The melodious voice of the call to prayer sounds to the unaccustomed traveler like a lonely man singing but is in fact the recitation of passages from the Koran spoken over a loudspeaker and in spite of my jet lag I was still able to recall my appointment with Tahany later in the afternoon. But for the moment I was all too happy to have the better part of a day free to wander around in Twain's footsteps through an ancient city, my first, a place I had only seen pictures of but never dreamed I'd be standing in just a short week ago.

Mustafa had suggested I leave the Egyptian Museum for later as Tahany had a surprise in store, so with 21st century guidebook in hand and what was left of my complementary breakfast rolled up in a napkin in the other, I opened the door to my Funduq and walked out onto the sidewalk, experiencing Cairo's sights, sounds, and smells for the first time in full daylight. Oddly enough, what caught my attention first wasn't the traffic as I had been warned so often of by the guidebook, but the peacefulness of the Nile, flowing in silent contrast to all the human activity around it.

Bazaars are what most often come to mind when thinking about the Middle East and so the Khan al-Khalili, described in my book as "an oriental bazaar of fable not to be missed", had easily made my to-do list. Everything else I came across, including what Tahany and I would see at the museum, I reasoned, would be icing on the cake.

I walked over to the sidewalk's edge, staring down at what was a more swiftly flowing river than I had first realized. Like all big rivers, the Nile near its shore appears alive and animated and yet towards its center almost eternal, which is one of the things I've always admired about rivers: they seem so preoccupied. In 1867, Twain couldn't help drawing comparisons either, writing that in Cairo "the Nile at this point is muddy, swift, and turbid, and does not lack a great deal of being as wide as the Mississippi."

A few feet away a vendor had made his own pyramid, stacking a few dozen oranges atop a folding table in front of a juice stall on the sidewalk, an older gentleman apparently intent on selling drinks, which I now felt obliged to investigate. Reaching into my blue jeans I emerged with a wad of Egyptian bills hastily exchanged back at the hotel. Egypt never got around to using money for nearly all of its long history, perhaps not so hard to imagine in this day and age of plastic credit cards, PayPal, electronic banking. At various times the brewer's yeast made all the currency anyone needed, staples like bread and beer, which could be exchanged for goods and services.[xxvii] [28] Modern Egyptian currency is colorful, reminding me of play money and my first purchase was for a glass of juice – a peeled mango squeezed together alongside half a stalk of sugarcane split lengthwise and fed through a metal vice. After finishing it I handed the glass back and hailed a taxi, which was soon whisking me the mile or so south to Islamic Cairo.

I did know already – having learned it in graduate school – why Robert Koch was in Cairo in the late 1800's. The German doctor, having wrapped up his work on anthrax, was now hot on the trail of another microbe, this one responsible for the disease called cholera. Koch was eager to apply some of the same techniques for cultivating microbes he had been so successful with for *Bacillus anthracis*. With the opening of the Suez Canal, travel between India and Europe had become faster in the 1870's and cholera – always the opportunist – was able to make its way from Asia to Europe that much faster, hidden within the ballast water in the holds of ships. Koch's main competitor, Louis Pasteur, had sent his own French team to Cairo because he too was looking for the cholera microbe during that same outbreak 140 years ago.

My guidebook offered advice on the best way to cross a street in Cairo – the largest city in the world located in a desert – something I doubted Twain would have required. It suggested that I use other pedestrians as a human shield. An amazing thing books are though, I thought as I walked, no less so than the printing press used to produce them, developed by Gutenberg, an innovation that would help bring an end to Europe's Dark Ages; making books more widely accessible and in quantities that, for the first time in human history, allowed knowledge to outpace religious authority.[xxviii] It was Gutenberg who realized he could combine the high pressures of the winepress along with small metallic stamps useful for imprinting designs onto leather, thereby creating a practical way to transfer print onto paper. He also took advantage of Europe's newly invented oil-based inks (the Chinese had already invented movable type, but they used wooden or clay stamps).[29]

Construction of Islamic Cairo was ordered by a General Jawhar as soon as the planets were aligned properly, and even though they were designed with pedestrians in mind, the layout of the old city is still much as it was centuries ago, back when Cairo was an important stop along the Silk Road from China to Europe.[xxix] Caravans rested not far from where I found myself that morning, and in the 1300's the Bedouins came there looking for exotic

---

[28] The English word *salary* comes from *salt*, a reminder that salt was sometimes used as payment. / Beer, unlike wine, is made from grain and since grain is easier to store than grapes, beer could be made more often during the year.

[29] The guidebook called this part of the city "Islamic Cairo", while the tourist map I held from Tahany's brother preferred the term "Fatimid Cairo" for some reason.

goods, forming the nucleus for what would over the centuries grow into the Khan al-Khalili bazaar, my final stop before meeting Tahany at her university. Unfortunately for Cairo's merchants, in 1497 the Portuguese explorer Vasco de Gama rounded the southernmost tip of Africa, circumnavigating the continent and in the process relegating Cairo to another unnecessary stop along the old Silk Road, whereupon the city soon fell into a prolonged period of decline.

After leaving the taxi and asking yet again for directions to reassure myself I was walking in the right direction I rested at an outdoor food stall, ready for my first taste of Fuul – Egypt's national dish – consisting of small brown beans soaked overnight in water, then boiled and topped with olive oil, or as I would discover a few days later in Luxor, a fried egg worked well in the mornings too. By the end of my trip I was even having the vendor mash it all together and stuff it inside a piece of pita bread so I could enjoy it while exploring Cairo's endlessly fascinating side streets. The Fuul I had that morning would have been recognizable to any pharaoh in ancient times; in fact bread is about as popular now as it would have been when King Tut was commandeering his chariots. But bread was a staple just about everywhere, it seems. Even Meriwether Lewis described it in his journal after coming across some Native Americans making a kind of bread out of ground sunflower seeds.

I should confess that I had ulterior motives that morning while visiting Cairo's Citadel. My primary goal at the ancient fortress built by Saladin (but currently a military museum with three mosques) was not so much to learn the history of Cairo as to catch my first glimpse of the pyramids, which my guidebook assured me would be visible off in the distance. And as it turns out, Cairo is full of surprises for I soon discovered I didn't have to go all the way to Giza to touch them, either. The walls of the Citadel are themselves made from many of the stones taken from the pyramids in the 1100's at the instructions of that same General Saladin.

After reaching the Citadel I climbed the stairs – in hindsight perhaps a little too enthusiastically – full backpack resting on my lower spine, then once at the top I gazed across a broad horizon spiked by minarets, many in pairs appearing to guard the domed mosques which rival professional football stadiums in size back home.[30] With the help of a woman from Belgium, we soon located the pyramids off near the horizon, so small they could have been mistaken as a staircase carved out of desert floor.[31] The area beyond them was so sparse that a woodpecker flying over would have to pack its own lunch, as my grandmother would have said.

---

[30] One of them – the Al-Azhar mosque & university – was founded in AD 970.
[31] The largest workforce in history up until then – 10,000 to 20,000 – was assembled to build the Great Pyramid.

I rested a few minutes, this time lingering on a section of the guidebook that caught my attention back on the plane, a page explaining the business of antiquities. I looked out at the Nile while digesting what I had just read. If not for the river and its periodic inundations every year, all of Egypt would have remained a desert and the material making up the pyramids would have stayed an ancient seabed, just as the vast majority of the nation is resting on today. Instead – complements of the Nile – Egypt was to become the most stable civilization in human history...until oddly enough not a single person alive would be able to read or write hieroglyphics.

Periodic flooding in the delta is why ancient artifacts, especially wooden ones, tend to be a harder find here than in Upper Egypt. The ancient capital of Memphis is still under much of this mud, which is why information about Lower Egypt comes mainly from writings found near Luxor...introducing an obvious bias for Egyptologists to contend with. Contemporary farmers in the Delta sometimes plow artifacts right into the ground without realizing it.[32] The Upper Nile, in contrast, has gotten so little moisture over the centuries that sandals, baskets, and even paper have been found in excellent condition manufactured 500 years before the birth of Christ.

With the help of a sidebar I became adept at estimating the age of Cairo's buildings. As a rule, the less ornate and intricate the pattern on the exterior of a mosque, the older the mosque tends to be. In the 1300's, the stone domes were plain. But over time, zigzags and floral arabesques began to appear as ornamentation. Minarets, which rise from the city's haze like needles from a pincushion, first began dividing up the skyline around AD 673, eventually lending Cairo another nickname, this one the "City of a Thousand Minarets". [xxx]

After visiting the Al-Azhar mosque, the bazaar was, mercifully, only a short distance down a narrow alleyway and so it wasn't long before I was elbow-to-elbow with locals thicker than cats at a fish fry, some barely old enough to play Little League back home yet capable of hoisting wide trays above their heads with cups of tea and flat bread piled high in the middle. I looked down to see one youngster too small to be carrying a tray walking beside me, looking up and smiling. It took another two blocks before I realized he had appointed himself my tour guide and another to learn that his name was Mohammad. His presence seemed to be having a deterring effect on the other vendors and for the first few minutes it's fun to be held in such high of regard, but after a while you begin to feel like a walking ATM machine in a place like this. In Twain's day, tourism was still so new that his party was

---

[32] Over the last two thousand years, the Nile Delta has covered entire cities like Damanur in silt, a site still being uncovered. / Some farmers, continuing a tradition begun in ancient times, still grind up nitrogen-rich bricks from ancient monuments in the Nile Delta as a fertilizer.

often mistaken by the locals as foreign agents intent on upsetting the carefully balanced power structure already in place for centuries. At other times they were accused of bringing cholera and inconveniently quarantined.

As Mohammad and I made our way down the increasingly wider alleyways towards all the activity, the air grew saturated with incense and carob, smells which gradually transitioned into eye-watering tobacco as the anthrax grant was fading more and more from my mind with each step.

My third purchase (yes, I was still keeping count) in Egypt was for two cups of tea, after which my guide and I pressed on. It was about then that I began getting the uneasy feeling I was being led by a 10-year-old. Mohammad always seemed to be ahead of me, turning around and waiting patiently whenever I was unable to shuffle sideways through the thick crowds due to my backpack. I couldn't slip through like he could and at times I imagined we were making our way through a dense jungle, except that, rather than vines blocking our path I had to contend with belly dancer costumes and colorful rugs slung just far enough overhead to require a duck.

I tried explaining to Mohammad with hand gestures what it was I was looking for, the kind of souvenir I wanted, delineating with my fingers in the air the outline of an Egyptian papyrus, but I still hadn't quite gotten the message across even though he nodded as if he understood me. At one point the boy even used his own sign language to enrich our tour.

Using his index fingers he made a sweeping motion across first one wrist and then the other. "Teiff....cot," he explained while looking up to see if I understood. Not until his second hand had been severed did I realize he was trying to tell me that the bazaar once had laws requiring the loss of a limb for stealing. We weaved back through the crowds for another ten minutes and just as I was beginning to lose confidence in him, we came upon a vegetable stand where I was instructed to "Come ein" by an older, graying gentleman pulling out a wooden stool, as if he'd been expecting me. He had materialized out of the shadows sporting a white beard, turban, and the same smile I'd come to expect in Cairo. It seemed the vegetable stand was performing double duty as an antiques shop. Grinning like a butcher's dog, he turned and patted the boy on the head, handing him something I didn't catch sight of, after which the boy disappeared into the alleyway before I had a chance to thank him.

"He is my son," the man explained, toothpick hanging limply from one corner of his mouth. He was speaking in hushed tones, as if comforting a small bird he was trying hard not to scare off its perch. "So tell me something. What brings you here this fine day if I may be so bold as to ask? Did you come alone my friend?"

Relieved to be able to speak in complete sentences again since the Belgian tourist, I explained how the papyrus I was looking for needed to be a genuine replica, not the kind Mustafa had warned me of, the kind where the paint flecks off every time you roll it up. I also didn't want one of the knockoffs made of palm fronds or bamboo leaves in place of genuine papyrus. The man seemed to be listening but unfortunately was no longer smiling. Undaunted I continued on, describing how I wouldn't settle for anything less an authentic scene copied from the walls of a temple or a tomb, preferably of ancient Egyptians baking bread or brewing beer. Unfortunately, his English wasn't fluent enough that I could explain why I needed it – that it was because I wanted something of the yeast to bring back home for inspiration.

His next move was to stage a disappearing act, venturing into a square room made of hanging rugs, after which he reemerged with a tray full of carefully balanced sheesha pipes, all copper and crowded together, each identical except in a bewildering assortment of sizes. Still stooping, he selected the pipe nearest me and pretended to take a puff from it, apparently deciding I needed to know how one worked. Feeling somewhat obliged I selected the smallest and feigned an interest, more out of politeness as I still had a dozen more days to fill up my pack, yet his son had been so helpful, if indeed Mohammad had really been his son.

"Do you have anything else?" I asked as I handed him back the pipe and stood to make my way over to the fruit stand, back towards the foot traffic, already putting into place my escape plan. His face drooped and his overall posture changed to disappointment. Then, just as quickly he produced a business card, explaining how there was money to be made if I could sell some of his fine pipes back in America. It was a game of chess and we both knew my turn was coming.

"How much for grapes?" I asked, plucking one off and holding it up to the light filtering in between the cloth awnings. Turning it over gently between my thumb and forefinger, I couldn't help admiring the dark, rich surface, as well as the white patches on its exterior, so white in places it stood in brilliant contrast to the purple of the skin. Except for wild honey, the grape is perhaps the best surface in nature for capturing the wild brewer's yeast.

The Egyptian took out a 1-pound note to indicate how much he expected and then, curiously, began reciting a verse, only part of which I caught through the roar of voices. "Drowsing through the climbing vine, the bee hints at the promise of sun-kissed wine."

I reached into my pocket as he went to get a plastic bag, giving me another chance to examine the grape. It was what I had come half way around the world for. The yeast was in my hand, a part of the grape's natural flora, awaiting its opportunity to turn the grape's sugars into more yeast. One of the things that makes life different from the rest of the natural world – a place of rocks and wind, mountains and rivers – is its ability to reproduce, and most microbes do it whenever, wherever, and however they can. I recalled one sermon back in Thistle about how, after Moses and his people had fled Egypt and while still wandering around in the Wilderness of Paran, the prophet had sent his scouts ahead to Canaan, after which they brought back symbols to show the land was fertile, one of those being a handful of grapes. There was no sign of disease anywhere on its surface.[33] I took my change and used what was left of my afternoon to hunt down the surprisingly elusive papyrus.

Papyri were the first portable books of the ancient world, the Ebers papyrus alone containing some 800 spells, one describing the proper use of opium to keep babies from crying, another on how malachite could be used to cover an open wound. Thanks to the scientific method we know today that malachite contains copper, an effective antiseptic for killing staphylococcus bacteria.[xxxi] The Egyptians had other uses for this mineral, sometimes mixing it with animal fat and using it as a green make-up. Other heavy metals like lead would have been applied around the eyes to darken them and, like copper, lead's antibacterial effects would have offered protection from infections initiated by the constant exposure to dry sandy winds.[34] If Mark Twain had visited Egypt during Roman times and his ship had anchored in the harbor of the world's first cosmopolitan city – Alexandria – the *Quaker City* would have been searched and its captain required to hand over every last papyrus on board so it could be copied and deposited in Alexandria's magnificent library.[xxxii]

As I made my way from the center of the bazaar back towards its perimeter where a taxi should be waiting, I encountered less of the alabaster pyramids and stuffed camels and yet still no papyrus. I was beginning to wonder if they existed. The guidebook promised if I walked north I could find just about anything, which I was sure I had. I was raised on a ranch so directions come second nature; the sun is great for telling direction when you have it. And yet the book hadn't let me down yet. Growing tired again, I found another bench, this time outside a coffee stall, and sat while reading in a sidebar about how the ancient Egyptians used beer as a solvent to suspend onions and salt in to induce vomiting if someone was unfortunate enough to be bitten by a snake and that Egypt has over thirty poisonous snakes, making the legend of Cleopatra using an asp to poison herself more likely than not

---

[33] Today, more grapes are grown than any other fruit. / Pliny the Elder described 91 varieties of grapes in the first century AD.

[34] Several heavy metals have antibacterial properties, and I was reminded of how Meriwether Lewis would administer a mercury salve to his men whenever they had bouts of venereal disease they acquired from the Mandan women. The use of this treatment inspired the 19th century adage "A night in the arms of Venus leads to a lifetime on Mercury." As a Montanan I grew up learning about Lewis & Clark and as an adolescent boy this bit of exotic information didn't escape my notice either.

true. Yes, I'd picked a good guidebook, I reassured myself while rubbing my lower back with my fingers to ease the growing pain. At least I'd had the foresight to do that.[xxxiii]

INVASION OF PRIVACY

In the rear of the cab on my way to American University, I couldn't help looking in the mirror again, smoothing my hair down while trying to remember what it was that I was doing, exactly. On the one hand, I was going at Tahany's invitation because she wanted me to know more about her work. But what if something more came of it? I squeezed the handle on the door tighter as the driver took yet another chance I wouldn't have, after which he slammed on the accelerator and we darted down what I took to be the garment district, catching random glimpses out the window of vendors trying to scratch out a living selling tennis shoes on blankets almost touching the tires as the taxi soared by.

I was an assistant professor coming up for a full committee vote...my last chance at tenure. I had signed a four-year contract when I joined the university; in other words I was a temporary hire. If I didn't get my tenure I'd be out of a job in a year. It was that simple. And if that wasn't enough, all of it was during a time when there were some alarming cutbacks taking place in science, unprecedented ones in fact. The Iraq War, Hurricane Katrina, deficit spending, Afghanistan, all had come together like the perfect storm after 9/11. I could barely support myself if I lost my job and had to go back to being a postdoctoral fellow in someone else's lab, let alone take care of a wife on the other side of the world.[35]

I took the guidebook out and plotted my position on the insert map with my pen, drawing a circle and then studying the details inscribed. Not bad for my first day on what was for me a new continent, I reassured myself. The book claimed the university was in the same direction as my hotel, yet nothing looked familiar, even as we rounded Tahrir Square, maybe because we came at it from the opposite direction. As we veered around a corner I caught sight of some paunch spilling out of an unfastened gap in my shirt, my bare belly jiggling as if dancing to the rhythms of the road's potholes. Maybe I'd walk back to the hotel afterwards, I thought, and besides, getting some exercise might be a good way to take in the city after dark. I considered how, back on the ranch I never had time to put on any extra weight. But then I went away

---

[35] In 1970 the average age of a scientist in the United States when awarded his or her first research grant was 32. By 2009 it would climb to 42 and in 2012 it was reported that, due to budgetary constraints, NIH was funding only 1 in 7 grant proposals: the lowest in its history.

to college and everything took on a more sedentary tone working in a lab twelve hours a day.

Once we got to the seven-story red brick building I was struck by how it could have been on any campus I'd seen before. A handful of students were standing around a fountain, chatting like young people do everywhere, discussing friends or upcoming tests most likely. Just inside the door on the wall was a directory and I scanned the names behind the glass. The sign above the door in front of me identified it as lab 123. Tahany's lab was 723. I looked up through the high stairwell, then down again at my gut. We had missed our first meeting and now she was directly above and the butterflies began again. On the way up the stairs my pack seemed heavier, which struck me as odd since I hadn't added anything to it.

After a few more minutes I had to take off the pack and sit down to rest. Tahany, like many MD's, was working on a project that had the potential to yield more immediate benefits than basic research like Ph.D.'s tend to work on. Ph.D.'s are usually more concerned with uncovering basic theories and mechanisms underlying a particular phenomenon, it's how we advance, get our work into print, gain tenure, etc, while MD's tend to be more focused on helping their patients, even if there's no particular explanation for how something works, a potential new drug perhaps, or a test to diagnose a disease faster and more accurately, as Tahany and her group were working on.

An important problem with schistosomiasis – Tahany's field – continues to be that the tiny worms causing the disease not only invade the body, bypassing its primary defense, the skin, to lay its miniscule eggs wherever it can once inside, but these eggs clog the circulatory system and set off inflammatory reactions in the process. Even worse, the worms can go undetected for years, thus aiding their silent journey back into the water supply. It seems a person can become infected and carry the worms without ever knowing it. Called *asymptomatic carriers* it's why one in ten Egyptians still harbors the worms.[36] In fact, worms cause more socioeconomic damage worldwide than any parasite other than the single-celled protozoan responsible for malaria.

The number of people living with these tiny worms is staggering, even to someone who studies diseases, something like 200 million globally, but no one knows for sure. The worms are so common in some villages that symptoms like blood in the urine (called hematuria) can be seen as a rite of passage for males entering adulthood, not so different than the way menstruation is for young women elsewhere. Tahany was trying to attack the problem of treatment selectively. Her idea, I seemed to recall, was that if a

---

[36] Typhoid Mary was a famous carrier of the *Salmonella* bacterium.

test could somehow identify non-symptomatic carriers of the worms, then mass treatment of entire villages with strong medicines like praziquantel could be made obsolete. All I knew as I struggled up what remained of the stairs was that it had something to do with finding new ways of detecting the elusive worms.

By the time I managed the 5th floor I had come to the conclusion that, what with teaching and grading, not to mention writing one too many grant proposals the previous semester, it had all finally caught up with me. Standing on that stair half out of breath meant there was no longer any denying it; I had officially entered middle age. The same piece of luggage I had been so proud of just a few hours ago was no longer a source of convenience but had at some point turned on me, intent now on carving its way ever deeper into my shoulder blade. The pain was excruciating, burning even, as if someone had stuffed rocks into my bag while I wasn't looking. I sat down again to catch my breath, remembering how, on their way to the Pacific, Lewis & Clark would walk alongside the Missouri River for 30 miles or more…in one day! They were both about the same age I was now. Looking for any reason to stall longer, my eyes caught on one of the research posters hanging on the wall nearest me. May as well have a look, I allowed myself after all, early 19th century explorers didn't have to sit around grading papers all day.

With posters detailing work presented at past meetings, Tahany's department was like any I had seen before. The names were in Arabic, but everything else seemed to be in English, this being the world's default language when it comes to science.[xxxiv] Many of the posters were either about West Nile virus or Rift Valley Fever; only one was focused on cholera.[xxxv] Robert Koch would have been pleased to know his work hadn't gone unfinished, even if his name wasn't listed on any of the posters. Apparently, the 5th floor was where the virologists and at least one bacteriologist were located.

Back home we often hang onto our posters detailing work (used as visual aids at seminars) because it can lend a sense of accomplishment to an otherwise dull existence in academia. And besides, it beats just throwing them in the wastebasket. Despite what a lot of TV shows try to convince you of, doing science isn't what you might expect most days. [xxxvi] I often find myself walking out into the hallway of the dungeon just to have a look at my old posters whenever experiments go awry or take too long, usually late in the evening when negative results seem that much more ominous for some reason. Mustafa had mentioned on the ride from the airport what Tahany's budget was for her entire lab and I calculated that it probably amounted to less than what I paid one of my two technicians in a full year, making me all the more curious to see what her lab would look like. She must be pretty resourceful is what I remember thinking.

On the day I trudged up those stairs, I had been in research for three years as an independent investigator, and before that working as a postdoctoral fellow in someone else's lab to gain more experience, and even before that for seven long years as a graduate student, and I have yet to step foot inside a lab that actually impresses me, aesthetically, when I first walk in. Even the nicer ones aren't much to talk about, more like your uncle's basement, the one with the pool table that has become a catchall for every stray item in his house.

Research labs like mine (so-called "wet labs") can actually be pretty scruffy places, stale-smelling incubators that could use a good cleaning out, stationed next to refrigerators which substitute as benches for empty test tube racks on their tops, floors sticky here and there with dried broth containing fossilized footprints no one has had the time or inclination to mop up yet, the kind of places you're more likely to see a mouse than some model with spotless teeth and lab coat holding onto the latest digital pipette and clipboard. There might be screws lying around belonging to things you wouldn't dare touch because someone is probably planning to use them for something they were never intended for, columns for separating (i.e. purifying) proteins for instance, pipes that look like PVC plumbing underneath someone's kitchen sink, complete with lubricated fittings.

The first time I was given a tour of a research lab, I was pretty disappointed to be honest. Back then I was still expecting to see something impressive. Too many books and movies, I suppose. Most benches in laboratories are long tables no more pleasing to the eye than those of a commercial kitchen, but instead of mixing pasta or cutting up vegetables on they're for doing science. The challenge of modern molecular biology is mostly an intellectual one, and sometimes it can be aesthetically pleasing if you're lucky, but not as often as you might expect.

Hot, newly poured Petri dishes have a peculiar and unappetizing odor to them, maybe because of the volatile ammonia salts and amino acids in their hot broth, and they require a flat surface for the agar to solidify on as it cools, otherwise you won't have nice flat surfaces to separate your colonies of bacteria onto after the gel hardens. The excitement, if there is any, may come afterwards when you're in an office pouring over your data, scouring photomicrographs and you happen to notice some detail that didn't catch your eye all those other times you looked at the same data, something you somehow missed back when you were doing the experiment.

In graduate school, the task I didn't mind the most (I'm kind of ashamed to admit) was washing the lab's glassware. It was mostly physical, not mental, and it took no real effort and I was guaranteed of seeing progress, not so different than watching dirt pile up alongside a posthole for a new fence on a ranch. Sometimes a person needs to see progress, any kind, especially when a single experiment can last more than a week before you get any results to share – good or bad, but most of the time somewhere in between – meaning that you have to repeat the whole experiment all over again because the answer just wasn't clear enough this go around, changing only one of the many possible variables to see if that was the real problem or not.[37]

Tahany's lab would be no different than other labs I'd seen except that it was smaller, with just two rows of benches dividing it up into three claustrophobic workstations. Some of the desks were built right into the benches, making it look at first glance more like a library. There were books and catalogues piled high as if waiting for the slightest tremor to set off an avalanche. I cupped my eyes to reduce the glare from the hall as I looked n through the window, my attention settling upon two figures, a man and a woman about the same age, perhaps 30 or so. The woman was taking samples from a rack of test tubes using a slender glass pipette while holding each up to the light and I wondered if this exotic creature with the dark hair gathered up into a bun could be Tahany. I had seen photos of her and yet I still couldn't be sure. Her hair had always been down around her shoulders in the photos.

My heart cheated me out of a beat or two as I twisted the knob and then pushed on the door just enough to creak open. Then I stuck my head inside, feeling a bit apprehensive remembering that Tahany still didn't know what I looked like. Like I said, I've never taken a photo I've ever liked. The young woman turned and smiled, pipette still in hand. I was beginning to think everyone in Egypt smiled; maybe it was a defense against the heat.

"Right on time, Benjamin. But you look so tired. Come in...sit down...please," Tahany offered up a chair that had some books on it as she finished transferring what was left of the yellow broth into an empty test tube. "So how was your first day? What do you think of Egypt so far?" Her tone seemed to match those of her e-mails, and I found myself relaxing a bit.

"Fine," I answered, "Having a wonderful time. Still no papyrus, though. And I've had a heck of a time finding any sunscreen over SPF 8. Who would have thought? Egypt being in a desert, I mean." That's great, I thought. Now I must be sounding like an old fussbudget to her.

Tahany indicated with her shoulder where she wanted me put the books, on top of what turned out to be a desk. The giveaway was the ceramic cup performing double duty as a paperweight, necessary due to the twin ceiling fans whirling away above our heads.

---

[37] I've often felt Meriwether Lewis would have been a better scientist than me. As Jefferson described him in a letter once "...of sound understanding and a fidelity to truth so scrupulous that whatever he should report would be as certain as if seen by ourselves..."

"This is one of my students.   His name is Kenan.   He's from Turkey."
American University has an eclectic group of students and faculty I would
come to find out that day; a reflection of the Cairo I'd seen since arriving, little
different than New York or London is that way I suppose.   "Benjamin is from
Montana."

"The States?" Kenan asked, eyes widening.   He asked it with the same
tone I'd heard everywhere that day whenever someone wanted to know
where I was from whether it was in the bazaar or walking down one of the
side streets, or even sitting in the back of a cab.   I had been surprised to find
on this, my first day in Egypt that, even in the capital of the Arab world,
Americans are not hated like I'd believed for some reason, quite the opposite
really.   America is still viewed as a shining city on a hill by many in the Middle
East, a place always just over the horizon.   It reminded me of the way
someone from Iowa or Kansas who has never been to California might look
forward to visiting Hollywood someday.

I laid my pack on the floor, sliding it with my foot up against the bench, yet
for a brief moment unable to budge it; then consoled myself with the thought
that Mark Twain had the use of a donkey to carry his belongings around while
in Cairo.

Tahany's complexion was as smooth as corn silk, giving away the fact that
she spent the bulk of her time indoors.   It was as if her skin should have been
darker, like her hair and eyes, yet it was closer to the shade of milk.   She was
prettier than a tub of freshly churned butter.   Still apparently avoiding my
gaze, Tahany scooted the stack of manuals over to one side with her arm but
without losing any of the liquid broth in her test tubes.

"Are you prepared to learn about Bilharzia? I hope so.   It's about all we
think of in here these days," she wiped some beads of sweat from her
forehead with her sleeve while I thought about how I'd seen Egyptian women
taking pains to avoid the sun all afternoon, perhaps to look more like Tahany.
Her accent was noticeable, but not enough to hinder any conversation, and I
stopped even being aware of it at some point.   I could tell she was taking
pains to pronounce her words carefully and I flattered myself into believing
she was doing it to impress me.   The more she spoke, the more obvious it
was she had gone to medical school in England.   It was then that it finally
dawned on me I was being expected to shift gears onto a disease I was
unfamiliar with, wishing I'd taken more time to read up during the long plane
ride.   Bilharzia, I did know, was another name for schistosomaisis. Some
others in use depending on where you live are Katayma fever, swimmer's itch,
snail fever, and blood fluke.

"Actually, I've been ready," I lied, then added, "Like coming full circle."

Tahany's work with the schistosome worm and its lifecycle had been what
initiated our long string of e-mails my first year at the university.   That

morning back in Thistle I had been lecturing my "sunrise clinical microbiology class", a small group of pre-med students who met out towards the edge of campus, and I remember it so well because those mornings always began early enough that the frost was still on the grass and the student's invariably picked it up on their tennis shoes, causing them to squeak on the floors of the hallway and, like a deer caught in the headlights, I made the mistake of taking an obscure question from one of my more wide-awake students in the back of the room. Every morning class has one, except that they usually sit closer to the front. Well, having learned in graduate school that, with college students it's always best to admit when you don't know something because this also seems to increase their respect for you, I answered truthfully and got out of it by promising I would get back to him the next day. As soon as I made it to my office I flicked the computer on and began searching for an expert, anyone who could answer his question, one that had been about vaccinations. I fired off perhaps a dozen e-mails and Tahany was the only one gracious enough to return hers, which explains how I came to be sitting in her lab at that moment looking rather perplexed.

So Tahany knew I already had *some* knowledge of Bilharzia but apparently wanted me to know more so she could explain her project in detail, the one she and Kenan were no doubt working on when I wandered in. In some ways, I got the feeling she was explaining it as much to him as to me. Maybe Kenan had just joined and needed to hear the lecture again, to reinforce it in his mind. I often do that with my own students; reinforce things every chance I get. A lot of details in science need to be explained in different ways before they stick. But once they do, you often get a good student out of the deal.

"The schistosome worm even when fully grown, is barely visible without a magnifying glass so you're probably getting some idea how small her eggs must be. Here Benjamin. Can you find one?" she smiled again but still avoiding my gaze, picking up a glass bottle filled with clear alcohol that looked like water, only the sharp aroma giving away the liquid's identity. Tahany put the bottle and a magnifying glass down on the bench beside me. I picked them up and began searching for her eggs. I quickly reasoned that the worm and her eggs must have settled to the bottom, water and protein being denser than alcohol, so I began by looking for them there, using just the magnifying glass. Kenan set a cup of steaming hot tea down in front of me, then without turning due to the lack of space, retreated back the way he came.

"Bilharzia is an ancient disease," Tahany began, "in fact my ancestors never did get the connection...that it was caused by a worm, probably because of their small size, but also because, even in ancient times, with their respect for the dead, cutting up bodies was actually frowned upon. Bilharzia was only known about by the symptoms it caused. The worms themselves weren't even seen until the 1850's. That's when Theodor Bilharz, a professor of anatomy, saw them during an autopsy at a medical school here in Cairo."

As Tahany talked, Kenan began pointing towards an old portrait, a black and white photo on the wall at the far end of the lab just above the door. At first glance, Bilharz looked like one of those Civil War generals with his flowing white beard, except that rather than a uniform Bilharz had on a tuxedo.

"In 1910 the calcified eggs of the worm were noticed by the careful eye of Sir Armand Ruffer[38] while dissecting the kidneys of two 20th dynasty Egyptian mummies.[39] I sometimes wonder what my people would have thought if they had known the same medicine – antimony disulfide – would remain the most effective treatment for schistosomiasis over the next three thousand years." xxxvii

I set the magnifying glass down, my way of admitting defeat, still unable to find any of the worms even after letting them settle to the bottom where they should have been easier to spot. Maybe it was the heat I consoled myself. The only time we get any real heat and humidity in western Montana is for about a month in the middle of summer.

"Antimony is a heavy metal," I explained as confidently as I could in an effort to keep the conversation from drifting into awkwardness.

"It's not easy," Kenan reassured me as he plucked one of the worms out with a pair of tweezers and, due to the lab's confinement, handed them off to his boss, who in turn handed them to me. Apparently, I was getting another chance.

"You see? They're so small. But the real problem is that they are such prodigious egg layers." I still couldn't see the worm so I took off my glasses. Being nearsighted I use them mainly for distance. It was then that we finally came face to face. The worm was extremely tiny, narrow, and snow-white. It was no wonder I couldn't see any of her eggs. They were microscopic.

"Different types of worms have their own preferences as to which areas of the body they invade and within a single mated pair hidden within the walls of the bladder, she may lay hundreds or even thousands of eggs in a single day," I nodded as Tahany spoke into my ear. It was the closest we had been and I could now smell her perfume. It's a strange sensation having the scent of someone you've just met up inside your nose, kind of like a pleasant invasion of privacy. Tahany was more beautiful than her photo let on.

"Well, the real problem isn't her size, but the fact that only about half her eggs will ever find their way back out...to be excreted. The rest will get

---

[38] A pioneer in the field of what would someday be called *paleopathology*. Ruffer was a professor of bacteriology at Cairo Medical School and he also identified TB, atherosclerosis, and various dental conditions in mummies.

[39] The 20th dynasty was during the New Kingdom period, about 1200 BC.

shunted...build up inside the vital organs, like the liver or the brain, triggering an immune response," she sighed and then forged ahead using a more resigned tone, "The body can do more harm than if it hadn't done anything at all, had simply ignored her eggs."

"Isn't someone in your department trying to find a better drug?" No sooner had I gotten the words out than Kenan, who had worked his way back around again, was removing the lid off a cardboard box and sliding a large, clunky, dissecting microscope in front of me. It was the same make of microscope I had used to see the yeast cells just a few days before. I looked through the twin eyeholes but could see only a few specks of dust, then felt foolish once it became clear Kenan hadn't loaded the slide into the bay and I wondered if Tahany realized what kind of a disheveling affect she was having on me.

"Oh, we're not looking for new drugs anymore," Tahany explained, "These days the emphasis is not on treatment but prevention. Some of us are trying to rid the snail that the schistosome worm spends part of its lifecycle in. Almost anyone can get bilharzia if they're not careful, even Napoleon's army got it. Backpackers and Peace Corp volunteers have sometimes picked up the worms just by using river water to shave with. You see...the adults can penetrate right through skin. They chew their way in by secreting enzymes, digesting proteins. These enzymes work quickly, too. The whole journey from water into someone's bloodstream can happen in minutes."

"Not the kind of souvenir you probably want to take back home," Kenan added with a grin.

"The greatest obstacles," Tahany explained, "are as ancient as Egypt herself; poverty and lack of knowledge," she frowned, "Eradication is difficult because the worms can hide inside of horses and dogs whenever humans are unavailable. The worm has one weak spot, though. It needs a snail. It has to spend part of its life inside this one snail. Not just any kind will do." Her words made me think about smallpox and how, even in this day and age of CRISPR and recombinant DNA technology we're all so proud of, how the smallpox virus is still the only human pathogen that's ever been completely eradicated from nature. Smallpox was exiled to the deep freeze mainly by vaccination and international cooperation, and of course money, but also because the virus had an Achilles heel: it could never hide inside another animal. Smallpox is strictly a human virus and so it infects only us.

Tahany continued: "The schistosome worm has relatives, over 300, inhabiting the bodies of a billion people at any given moment. Dams, irrigation projects, they all have their place in improving the standard of living here, yet they come with a price because they create new habitats for the snail. My country has used crayfish to eat the snails, which has met with some success. Yet nature always seems a step ahead. There is even some evidence the worms have evolved to take advantage of human activity, the times of day we bathe in rivers coincide with the times the worms release themselves from the snails."[xxxviii]

I looked over at Kenan, who seemed as if he might have been waiting for a break in the conversation. When he saw my looking at him he pushed the microscope closer. Loaded into the bay was a glass slide with a pinkish film smeared onto it. As I began my search, the medical student explained how I was looking at a tissue slice of a former patient's bladder. Clearly visible through the lens were the large, flat plates – skin cells being impossible to mistake – along with dozens of tiny pink specks, which turned out to be the eggs scattered throughout the tissue, some equipped with grappling barbs that looked like fishhooks. They had been subjected to a synthetic dye because nothing in nature is that bold of color, not even red blood cells are that brightly pink. Few things in nature's realm can match the loud colors that synthetic stains manufactured in a chemist's lab can. Next we examined a drop of urine from the same patient. Looking for the worm's eggs in the patient's urine under a microscope is how laboratory technicians still find them to make a schistosome diagnosis. It's tedious work, the kind that reminds me of trying to coax a hot mule out of a cool barn, the kind of work I'm glad I don't have to do much of these days.

The eggs in the urine were few and far between. "They seem rather sparse and difficult to find," I said. "Which must be why we have graduate students." I was relieved when Tahany smiled. Her eyes caught mine for a moment and then, like a butterfly, flitted away again. Maybe she wasn't so put off by my appearance, I reasoned. We seemed to be hitting it off. If she was disappointed in my physique, at least she was kind enough not to mention it.

"And this sample is from a patient who shed a lot of eggs."

"There must be a better way," I added, "Didn't you say something about DNA? Weren't you...?"

"Yes. I was just coming to that, Benjamin. We've been developing a PCR test."

"So you're doing an amplification, then?" [xxxix]

"The same technology that has allowed the sequencing of DNA from ancient mummies works well for worm eggs in a patient's urine."

"So you want to use PCR to sequence the worm's DNA?" I ventured.

"No. Not sequence it. Just detect its presence."

"We only need the DNA from one of her eggs," Kenan explained.

Tahany finished his point for him, "In theory, only one egg is needed to confirm the worm's presence, and since all her eggs contain the same DNA then...", there was a brief silence as she considered her words. "That's the reason PCR works so well for crime scenes and why you only need a drop of blood, and if you can amplify it, that cell's worth of DNA means you can have as much as you need, even for DNA fingerprinting and sequencing if you care to do it."

Tahany reached above her head to the shelf and opened one of her notebooks. A dozen or so textbook-sized sheets of film cascaded onto the bench. They were large negatives with images of what looked like ladder rungs but without any railings. I knew that each rung on the negative represented a different length of DNA fragment amplified enough times to make the spot visible on the film. The worm's DNA was discernable because, during one of the steps of PCR, all the fragments were labeled with a radioactive molecule, a probe, tagging and identifying the DNA like a lighthouse can a shoal, labeling it as coming from the worm. PCR is a bit like amplifying a page from a book simply by making Xerox copies of it, and then making copies from the copies using an additional machine and so on, which is where the amplification part comes in. PCR is all about making DNA copies of very specific pieces of DNA, just like when you want to make a copy from a book you select only the pages you need, not the whole book. When you use a PCR test to probe an organism's genome, you usually look at only one of its unique genes to identify it.

"This lane," she ran her fingers down the surface of the film, "it shows up fairly well. Don't you agree Benjamin? But it's not as reproducible as we need it to be. This is DNA from one single egg cell. But sometimes the signal blends in with the background and that's the problem now."

"Noise is always a problem. In every experiment," I sympathized.

"When you can see the worm DNA, it amounts to a red flag, a tag identifying a patient who's been infected, someone who can now be put on the drug and treated. No more shedding of eggs back into the environment." Tahany must have noticed I was impressed because she next tried to temper my growing enthusiasm, "There are some problems, though."

"Like what?" I demanded, "These results look really promising, Tahany. What's the matter?"

"Like the background noise on this negative over here," she ran her fingers across one of the fainter rungs on the negative, "and how sometimes the test says someone is infected when they are not really. Perhaps these are rare instances of cross-contamination. Or maybe the patient is infected…but just not producing any antibodies yet. We don't know. It takes only a single egg to contaminate an entire sample…and if it's an egg in the wrong place in the lab…and I forgot to mention how these eggs do like to stick to things…then your test can lose all credibility."

Kenan, who it was obvious now wasn't new to the project, added, "While we believe the test can work, we are still left with the same old regulations

The government wants to cling to the standards for diagnosis set in place decades ago."

"And in the meantime," Tahany added, "others will become infected. More work needs to be done to convince the authorities. But I just don't know how…" she looked away from me. "Things move so slowly around here. It's probably not like that where you come from, is it Benjamin?"

"But it will work," I was unable to answer her because parasitology and government regulations were out of my area.

We continued looking at the negatives and then studied the statistics her group had compiled from the data, checking and rechecking the numbers, until we were interrupted by a door slamming shut. We all looked up in time to see a small, slender, well-dressed man in three-piece suit entering the lab. The closer he got, the older he looked, and his salt-and-pepper hair was receding noticeably at the temples. I looked at Tahany who was now smiling.

"Oh, I almost forgot. The surprise. This is Gamal. Gamal is a microbiologist. Like you. I asked him to show you around the museum tomorrow."

We shook hands as I tried to hide any disappointment that might have been obvious. I'd assumed Tahany would be my guide. A slight, bespectacled man with soft hands grasped mine…definitely the hands of a scientist. He was older than Tahany, perhaps 50, I guessed. His English would turn out to be as capable and yet his delivery lacked the same authority Tahany's had.

"Gamal works in the department, part time with me," she explained, while I resisted the temptation to look around her lab to see where it was he kept his belongings.

"But I mainly specialize in the brewer's yeast of which I hear you are quite interested at the moment," Gamal said.

"Gamal is an expert. He works at the Stella brewery," Tahany added, apparently feeling the need to give him a reference, "He knows all about yeast genetics."

I had already been surprised on this, my first full day in Egypt, to find not only toothpaste, but beer and wine available in grocery stores, but perhaps I shouldn't have been. Ten percent of Egypt isn't even Moslem…it's Christian. In fact, the coming of Christianity after the crucifixion of Jesus is the reason Egyptians stopped mummifying their dead. These strange newcomers insisted that the body was a shell meant for carrying around the soul, a soul that ascends to heaven after leaving behind something no longer needed.[xl] And as I learned that first day, many Moslems in Egypt drink alcohol, even though the Koran forbids it. It's because in life few things are as simple as you might first imagine and much the same way many Christians don't necessarily feel the need to go to church every Sunday to be good Christians, there are Moslems in Egypt who will allow themselves a glass of wine or a beer occasionally.   It's near where it all began after all, brewing and winemaking.   In fact, Egypt has no less than two major breweries in addition to a domestic winemaking industry.

"And you're in luck. He's also an Egyptologist," Tahany said with obvious pride.   As for Gamal, he didn't say much of anything, contenting himself with studying me.   I was studying him too, but trying not to be as obvious, wondering if he and Tahany were only coworkers, or whether there was something more between them.   There was affection in her voice and I could see the two of them together, if not physically, then at least in an intellectual sense.

"It's not luck, really," Gamal added, "Most Egyptians are Egyptologists " Everyone laughed except for Kenan and I got the impression he would have preferred returning to his work.   From the look of things on his desk, he had a lot to do before calling it a night.   I started looking around for my backpack, hoping my search would be taken as a sign I was getting tired but Gamal was still curious and wanted to know more about me.

"So what made you become a microbiologist? Everyone has a story, it seems. More often than not I find it quite interesting...if you don't mind my asking."

I had my stock answer prepared beforehand, before I even climbed the stairs.   And I delivered it the same way I always do.   "I was raised on a ranch, but for some reason never wanted to be in the rodeo," I began, "Maybe because I'm not all that tall...at least by Montana standards.   Instead, I just left the roping and calving up to my brothers.   I was more interested in finding out what was living inside the bull than riding up on top of him."

Tahany smiled and I could see my story was having its intended effect of lightening the atmosphere while deflecting some of Gamal's scrutiny from me, except that Gamal hadn't changed his expression much, in fact it seemed as if he was hanging on my every word, weighing each one carefully to find which might have the hidden clue inside.   God, I thought to myself, he sure is the studious one.   And I also thought about how I had learned over the years that people as they get older tend to lose the rough edges that once defined them as a child.   Like a rock in a mountain stream that slowly becomes rounder, approaching perfection given enough of a chance, as an adult I had gotten better at evening out my own edges, the ones I didn't want anyone to notice, as well as turning something sad into something not quite as bad as you might have first imagined.

I rarely let on, even to those I knew well, just why it was I had been drawn into the strange world of microbes and it's kind of a mystery to me even to this day how a single afternoon can stand out as if to the exclusion of all the others.   One day after having recently turned thirteen, my family's prize steer

– the one my father was planning on breeding – suddenly rolled down into the bottom of a ditch beside the road, as if toppled by a strong breeze. I happened to be the one to see it while on my way home from school.

Our family's ranch is situated where the nearby mountains transition into foothills as they come down to meet the Lewis Valley below. Our valley is broad and flat for the most part and named after the explorer Meriwether Lewis, one of the leaders of the Corps of Discovery. The town of Thistle, where the university is and where I had hopes of gaining my tenure, is in the same valley but at the opposite end. Thistle's one of those towns in the American West that remind me of charms from a bracelet that somehow manages to dislodge and roll, not behind a refrigerator or a stove, but between four mountain ranges, the kind of place where farmers follow the river down into town every Sunday in their pickup trucks to go to church and stock up on supplies while catching up on the latest gossip as they get their hair cut in the town's only barber shop, not the fancy hair salons the tourists on their way down to Yellowstone use, and they wear plain blue jeans without any extra rivets in them, the kind with the straight boot legs that hug the lower portions of their body so tightly, like saran wrap, that they have to walk a bit sideways no matter what the style happens to be in the big cities.

Brucellosis was the vet's pronouncement, which sounded like a death sentence. And it is...at least for the animals. Brucellosis is caused by a tiny bacterium, one so small, even by bacterial standards, a pest new to North America[40] and of course so tiny I would never see it on that day, or any day for a long time to come, not until I entered my second year of graduate school, and yet I was still struck by the ferocity of the thing, the rapidity of something so small as to be invisible, and by the fact that it could bring down something so many times larger than even my grandfather was. I was completely fascinated by how quickly this once noble beast had been rendered helpless by something so tiny. And to a 13-year-old boy with an overactive imagination, the whole thing probably seemed like the extinction of the dinosaurs must have been.

After the vet left, my father realized it probably began back when the herd's milk began drying up a month or so earlier, which meant my father had missed it, failed to recognize something every rancher is brought up knowing about, the early signs of brucellosis...a gradual decrease in the herd's milk production.

"Must have gotten it from a bison or an elk lookin' around for food," is how the vet explained it as he scuffed a small hole into the dirt road with the heel

---

[40] It was brought west from Europe. The bacterium makes an attractive bioweapon (like anthrax) because it can remain alive outside the host and can even be aerosolized (spread through the air). It replicates only after gaining entry into the host's cell – a macrophage – and is why once a cow gets the microbe, the animal either dies or it stays with it for life.

of his boot, trying hard not to look my father in the eye, and he was right, it had been a harsh winter, worse than usual, with plenty of snow on the ground, which meant not as much exposed vegetation. The vet's younger brother and my father had gone to school together as boys.

"Everyone has to eat, even bacteria," he muttered in that slightly ashamed tone people often use in the valley when they don't know what else to say. It was all clinical the way the vet and my father talked, but after he left that day I was sure of what I wanted to do with the rest of my life. I only explained to Gamal and the others in Tahany's lab how my parents had decided at some point that I was too different from my brothers, and that college might be a better choice, but I've always thought it was when I refused to chew tobacco that guaranteed me my tuition money. Something else I didn't mention to the group that evening was how aware I had been that my father suffered in silence a full two years after we had to cull the rest of the herd and my mother always said I got most of my traits from him anyway. I guess finding out ones parents aren't perfect, as is inevitable at some point in every young person's life, is difficult and yet strangely liberating at the same time, for it frees one from the responsibility of being perfect too.

Gamal offered to show me around the department, which meant an interesting, if sore, hour more of retracing our steps up and down the staircase, Gamal discussing the various projects being conducted at the school, periodically taking time to summarize the more interesting posters that caught my eye along the walls. I knew from my own experience that each poster represented the culmination of several months – if not years – of tedious work done by a team of scientists and technicians. Whoever said genius is 99% perspiration and 1% inspiration may have lived before the invention of automatic pipettes and disposable plastic Petri dishes, but he was still right about microbiology in the 21st century.

We wound up outside Tahany's lab as darkness was bringing with it the vulnerability one often feels upon finding oneself in an unfamiliar place as nighttime arrives. Just as I was getting ready to leave Tahany became more serious all of a sudden, guiding me aside by my shirtsleeve.

"Did you work out a time to meet tomorrow?" she whispered.

"Around nine," I whispered back but without knowing why, "The King Tut exhibit is supposed to be less crowded in the afternoon. His earthly goods take up a whole wing of the museum it seems."

"You don't have to worry about him being there," Tahany explained while looking intently into my eyes, "Gamal is my fiancé."

I'm pretty sure my heart cheated me out of another beat as she dropped the f-word, and I desperately hoped it didn't register on my face.

"Oh I'm not too worried," I stammered back, "Really, Tahany. It's been an amazing trip so far. If it ended tomorrow I'd say it was well worth it." I attempted a smile, still not sure if I succeeded, which made my way down the stairs an interesting contrast to my trip up. This time the trip left me winded not so much physically as emotionally. And I was also thinking about how if Tahany's work could be realized in her small lab on the top floor of a plain brick building in the middle of Cairo, how it would deliver so many benefits far beyond even Egypt. Bilharzia, I knew, is a problem in so many places around the world.

I gently refused Kenan's offer of a ride as it was only a few blocks back to the hotel, at least according to the guidebook, and I even managed it without too much delay and without feeling too sorry for myself, either. As I relaxed on the balcony back at the Bedouin, looking out over the three-story tiled roofs, I tried picturing what 1870's Paris must have been like, my stocking feet propped up on the railing, muscles still tingling, then wondered if it was offensive should anyone happen by, showing your bare feet that way. I changed my position to be on the safe side, considering how I'd forgotten to ask the most important question, yet here I was trained to ask questions...all sorts of them. But not those kind apparently. There are no classes in relationships in graduate school and I had managed to get blindsided yet again. I sure was a good one for wishful thinking. She was a catch and Gamal had to know it.

Strangely enough, I wasn't as jealous as I might have been in my younger days...progress of an unexpected sort. No, I was more resigned to being alone than I ever had before.[xli] I leaned further back, consoling myself with the notion that perhaps my own disappointment had less to do with Tahany and more with my own expectations.

One of the reasons I bought the ticket to Egypt in the first place was because of the book I'd come across as a teenager. When he boarded the *Quaker City* in New York, like me Twain was also a bachelor in his 30's. And while on the trip he fell in love, not so much with a woman as with her image. A fellow passenger, Charles Langdon, had brought onboard a photograph of his sister and Twain later claimed that after seeing Olivia Langdon's photo he knew right away they would be married someday. He'd already decided before they'd met. How's that for confidence, I thought? And when they did go out together for the first time, it was to hear another writer, one who had come all the way from London to speak. They would marry two years later, with the first royalty check for *The Innocents Abroad* arriving the same day and the book would soon make Twain famous.[41] I've often wondered whether they shared fond memories of their first encounter the night they heard Charles Dickens speak on his American Tour.

---

[41] *Innocents Abroad* became an American bestseller, with numbers approaching that of the Bible. Twain's book was sold by door-to-door salesmen as a subscription, often by Civil War veterans with limbs missing.

CHRONOLOGICAL ORDER

Blaming it on the jet lag, I awoke with the sun already having beaten me, which will always be a point of contention to a rancher's son no matter which hemisphere he hangs his hat in. I hadn't forgotten to ask for a wakeup call, even going so far as to give the attendant some baksheesh, then remembered that my phone didn't work. I must have slept through the call to prayer too.[42] It was 8:25, meaning I had just enough time to get dressed and get my sunburned, muscle cramped self to the museum. Unlike that first day, I'd be leaving my backpack behind. If a microbiologist could take time away from his job at the brewery to tell me about the yeast, then the least I could do was to be on time.

A road twice traveled is never as long, so it was a liberating jaunt down to the square without my backpack on. When I got to the same vendor and after buying another glass of juice, I turned north. From there, the top of the rectangular two-story museum was within sight. I bypassed the long line of buses and headed for the front gate. Not surprisingly, Gamal was already waiting. I looked at my watch – 9 o'clock on the dot – not bad considering I was ten time zones away from my comfort zone.

"So what do you think of her, Ben? Is she old enough for you?" Gamal was craning his neck to look up at the building's roof. He barely came to my shoulders. "However it depends on your perspective. When this building was first constructed, King Tut was still slumbering in his tomb...just as he had been for 3,000 years. It holds more than one hundred and twenty thousand relics. Just on King Tut's body Howard Carter found one hundred and forty items." Dusty and crammed and "a bit like going through someone's attic" was how my guidebook described the place.

Gamal used his pass giving us free access through the guide's door, bypassing a turnstile with a lone guard reading a newspaper. If he had looked up he would have caught sight of an amusing pair, the slight, dark, bespectacled Egyptian in strict three-piece suit without a wrinkle, walking next to the lumbering American in T-shirt, sunburned neck, cowboy boots and jeans.

"We will save the Royal Mummies for later," Gamal explained. "They're in a room all to themselves."

"The only mummies I've seen so far have been on the paper money."

---

[42] I later learned that the call to prayer used to be performed by a gifted caller, or muezzin, but because of Cairo's roaring, nonstop 21st century traffic, the *summon* has since been replaced by a tape recording over a loudspeaker, which is what I probably heard.

"You know, back in the 1800's, when Egypt's railroads were being built, mummies were in such plentiful supply that the engineers would use them as kindling for the boilers…feeding them in like coal."

I couldn't suppress a smile because I knew it already, then told Gamal about Mark Twain once claiming that, while in Egypt he overheard an engineer yell out to one of his assistants, "Damn these plebeians, they don't burn worth a cent, pass out a king." [xlii] I couldn't tell if Gamal got the joke or not because his expression hadn't changed. He did strike me as the serious one, even more so than the previous evening. Something was bothering him and I couldn't help wondering if it had to do with Tahany asking him this favor.

"Didn't they grind up mummies for medicine?" I asked in an effort to keep the conversation going. Freshman history was coming back to me in dribs and drabs. Museums are good at reminding me of things I thought I'd forgotten all about and as we entered I caught a whiff of the musty smell they all carry. Fungi and other hidden decomposers were busy spewing out gaseous products.

"Yes. Europeans believed that mummies could have magical, restorative powers…because they are so old. Even Queen Victoria took a bit of ground up mummy with her afternoon tea…or so I have heard.[43] Maybe it is not so surprising in an age when patent medicines might be laced with opium, that people would turn to mummies for cures." Peering up over the black rims of his glasses Gamal added, "Because the Tut exhibit will be crowded this morning, we need to begin our tour on the first floor. These rooms are all arranged in chronological order, which means we will be visiting 42 containing thousands of years of history beginning with the earliest relics from pre-dynastic times – some over 5,000 years old…older than writing itself – and ending in the more recent Greco-Roman period."

I thought about how most great museums in Europe, if they were lucky, only *begin* with the Greco-Roman period. This museum actually ended in it, as if ancient Rome were somehow modern.

"You know," Gamal explained, "civilized people have always enjoyed museums…and libraries. If not for the Library at Alexandria, we would never have known of a poet named Homer, or the playwright Sophocles. In Egypt is where the tradition got started…of copying manuscripts for posterity."[44] I nodded while recalling that exactly one week ago I was grading papers and so was happy to be anywhere else. Just walking inside the atrium was worth the price of admission…if I'd paid one.

"Did you stop to consider why we are on the east side? And your hotel too? It's no coincidence, you know." I hadn't, and admitted as much. "It is

---

[43] She also was a beer-lover.
[44] Only Egypt with its dry, hot climate has preserved any of the works by Homer directly from antiquity, all of which are on papyri.

because the east is where the sun appears each morning. The east symbolizes rebirth...the side of the Nile most Egyptians went about their daily routines on. It is also why the pyramids are on the opposite side, to the west. The west is where the sun went to die each evening, return to the underworld...so the west has always represented death. You are visiting the pyramids tomorrow I understand?"

"Yes," I answered, slowly becoming aware that Gamal had given this talk before. He seemed comfortable with it, and yet I still couldn't shake the notion that something was weighing on him. Then again, maybe it was just his nature. Perhaps he was moody, I reasoned. I sometimes am in the mornings for reasons I can't always explain. And besides, when you've only known someone for a couple days, it's kind of hard to tell.

"Tomorrow evening you will be traveling to Luxor, Tahany says?"

"By train. Is that the best way? I've got nine days more, then it's back to assigning homework."

"You should be fine," he answered curtly.

We were at the beginning of a long atrium extending before us. Within it were various monuments, some reaching almost to the ceiling, most made of stone, including sphinxes, easily recognizable even to a casual observer. Others were smaller and in the shape of people, pharaohs I assumed, some assorted arches and pillars were also present, even a small obelisk helping to echo back 21st century voices. I was excited and we hadn't entered a single room yet.

"Zoomorphic," Gamal said as he pointed in the direction of the sphinxes, apparently noticing my being enamored by them.

"I didn't know you had sphinxes...with different animals on them I mean...other heads on a lion's body. Falcons and hawks, there's even one with a ram's head over there. Look. It's as if they're interchangeable."

Gamal then explained how they represented local deities and how these sphinxes once had important jobs: to guard the temples. It was my first surprise of what would morph into many that morning...finding out that not all sphinxes were sphinxes.[45]

Gamal pointed to an inscription from 2000 years before Christ – hieroglyphs chiseled into stone – and with one finger running across he translated it, "The mouth of a perfectly contented man is filled with beer. My ancestors had a sense of humor. They made up words that sounded like those the object made. For instance, they called wine erp, like a slurp when you drink wine. Cats they called miw. With donkeys they said eee-aww.

---

[45] It's likely that Queen Cleopatra never laid eyes on the Great Sphinx since it was already covered by sand when she began ruling over Egypt in the first century BC.

Tahany informs me you used to work with anthrax. That is caused by a bacterium, is it not?"

"Yes," I answered, somewhat hesitantly because it was the first time I'd thought about the anthrax grant since waking up. More unexpected progress.

"My ancestors had a hieroglyph to represent anthrax. As you probably know, it was one of the plagues described in the Old Testament. Anthrax attacked Egypt's cattle...and then our people." [xliii]

"Didn't King Tut have some wine in his tomb?" I asked in an effort to divert the conversation. Gamal didn't know about my recent setback with the grant and I didn't feel like explaining it.

"Yes. Carter found 26 amphorae in the boy-king's tomb after unsealing it. We even know how some of Tut's wines must have tasted. Four were labeled sweet...some others had been imported...all the way from Syria in fact. [46] The wine's vintages were stamped on the jars and provided the name of the winemaker. Ancient advertisement of a sort I suppose. We can only guess at what the colors would have been. Egypt being in a desert means the wine evaporated shortly after the king was laid to rest, but the writings of later Greek travelers inform us that Egyptian wine was mostly red." [xliv]

We continued walking the perimeter of the atrium, Gamal explaining in his formal way how scientists can be so certain ancient people made wines. It reminded me of the way researchers studying grizzly bears at Yellowstone National Park will try to learn about them indirectly if possible, by studying the tracks they leave, including hair follicles, when they brush up against the rough bark of a tree.[xlv] It's a lot easier, not to mention safer, to collect DNA specimens this way both for the bears and the people, than it is tranquilizing them. Likewise, when looking for signs of ancient wine[47] it is the chemical traces left behind in the jar, its molecular fingerprint so to speak, that one looks for rather than the wine itself. It turns out the DNA of yeast has been recovered and sequenced from the bottom of wine jars from ancient Egypt.

"Clay amphorae and jars are good at holding onto organic molecules. For thousands of years, even. There is also a clay[48] winemakers still add to their wines to remove impurities like proteins...so the wine does not turn cloudy. It works because clay is negative and the proteins in the acidic wines are positive in charge. Does this sound familiar?"

---

[46] Wine is perhaps the most complex food, flavor wise, thanks to the yeast's unique chemical contributions.

[47] Ancient beer can be identified on pottery thanks to *beerstone* – a calcium oxalate mineral left behind by precipitation from the beer – in the Zagros Mountains of the late 4th millennium.

[48] The clay is *bentonite*, named by a geologist in 1898. He found a large deposit of it near Fort Benton, Montana. During the letter attacks in 2001, it was incorrectly claimed by the media that anthrax spores in the letters had been prepared (i.e. weaponized) using bentonite.

"Chromatography," I answered confidently. It was all too obvious where he was going. [xlvi] As most folks with even a passing interest in science know, opposite charges attract each other. Negative particles attract positive ones and vice versa. It's the same force that holds electrons and protons together inside of atoms. Separation chemistry really can be that simple sometimes. [xlvii] Often, to separate one chemical from another in a modern lab – different proteins from a bacterium or a yeast for example – we use ground up clay, highly purified and loaded inside a glass tube because the clay particles bind selectively and, just as importantly, reversibly, to specific proteins as the proteins all pass through the clay from top to bottom. The implication is that ancient amphorae from the Middle East behaved like modern separation columns do today, a strange but fortunate coincidence for molecular archeologists. For the Egyptians, clay had more practical uses, though. The Egyptians coated their beer jars with clay during storage and it probably helped clarify the beer, freeing it from impurities, as well as preserving it.

"Tartic acid is a good chemical marker for grapes," Gamal pointed out, "One pigment in grapes is a small molecule called syringic acid. It also binds to clay. So tell me, to a bacteriologist, is the yeast beginning to sound more interesting?"

"I suppose," I answered, "But I'm still not sure if I have the enthusiasm it takes."

"Well you have been involved with much simpler organisms, ones that are much better understood than the yeast. But what if I told you that, if not for the yeast, we would not be where we are today?"

"How's that?"

"If not for fermentation, civilization wouldn't have begun when and where it did. What I am saying is that the first farmers settled down and raised crops not so much to eat as to drink fermented beverages, which are rich in calories...made from wheat or grapes and fermented into beer or wine by microbes. [xlviii] Imagine a hunter-gatherer experiencing alcohol for the first time quite by accident. A difficult existence full of uncertainty is all of a sudden transformed into a relaxing, mind-altering one, a drink that can be shared with others, with a unique taste and a belly full of calories as a bonus."

LIQUID BREAD

I told Gamal about how, as a boy back on the ranch, I used to watch cattle in the winter dig down to get the grain at the bottom of a pile, down where the oxygen was scarcer and the microbes more plentiful. These microbes were fermenting the grain into alcohol...creating their own beer...like tiny workers in a microbrewery.

"Even animals enjoy getting tipsy. It is quite possible primitive people watched and learned from the animals all around, cultivating plants they knew would turn to alcohol more easily. Birds prefer eating the rotting berries more than the sugar-filled ones. Have you ever seen a drunken blue jay? It is quite a site."

"Animals are indeed smarter sometimes. Which is probably why I've never seen a horse placing a bet on a person." Even though he seemed to be loosening up more, Gamal still didn't get my attempt at a joke as we turned the corner of the giant hall and came upon a glass case containing some clay pots. Whatever it was bothering him may have been loosening its grip as we ventured further into the atrium.

"Fermentation is a way of preserving food. It is the culinary art of allowing food to rot...but in a controlled way. If food decays only a little, this decay can actually be good because it helps preserve the food. Strange how that works, but in an age before refrigeration, fermentation would have been one of the important ways people had of keeping calories safe."

He explained how alcohol produced by yeast is not only an aphrodisiac but also an antiseptic, and why most mouthwashes today contain 20% alcohol. In fact, alcohol will kill the yeast too, eventually, which is the reason wines are seldom above 12%. The brewer's yeast will succumb to its own waste. True enough, waste can be deadly if allowed to build up. One of my earliest memories was watching the Apollo 13 astronauts on TV on their way back from the moon after the accident happened and how they nearly died because of the buildup of their own carbon dioxide they exhaled, which became concentrated inside the spacecraft. The yeast release carbon dioxide too when they break down their food.[49]

We walked some more until Gamal, ever the perfectionist, stopped suddenly in his tracks. "Oh, I nearly forgot," he exclaimed, "the slightly acidic nature of fermented beverages. The yeast also discourages bacteria because of the acid it makes, but perhaps you knew that already...about the inhibition of bacteria using acid."

---

[49] Many of the victims at Pompeii may have been killed not by lava but by carbon dioxide gas released from the volcano ahead of its pyroclastic surge.

"It's why vegetables like tomatoes are easier to store in cans," I added "They're high in acid."

Gamal continued: "So you see the wild brewer's yeast begat the first agricultural settlements, in places like Turkey and Syria, because they gave us the ability to store liquids for longer periods...years if necessary...very important when you're talking about people living closer together in large numbers for the first time and having to share food. As for the Nile valley, it was a lack of water that encouraged the people to settle initially. Before the drought, they lived scattered in what was then a much larger Delta, portions that have since become desert. Changes on the Earth caused these marshes to dry up, so the inhabitants had no choice but to move closer to the Nile, and thus closer to one another. Egypt came about by a sort of...um...not explosion. What is the word you have in English...the opposite of explosion...?"

"Implosion?" I ventured.

"Yes. Thank you. An implosion. People living so close together for the first time meant they needed to be cared for...become better organized. It s no coincidence that hieroglyphics were invented when people were so much more in contact. They needed to keep track of supplies. Beer was a food...for them liquid bread. A source of calories. Famine due to the fickleness of the Nile caused a great many problems my ancestors lived in fear of, and for good reason, as we shall see. The first written language was numbers, boring yes, but important nonetheless, for they needed to keep track of how much beer could be brewed from a given harvest. In fact, every step in brewing has been found depicted on the walls of a tomb in Egypt somewhere."

"How about wine?" I asked.

"Are you asking about the oldest evidence? If so, then traces of wine have been found using techniques like mass spectrometry,[xlix] from clay vessels dating to about 5400 BC in Iran.[l] I believe that is the oldest confirmation of wine we have. Every civilization has taken advantage of the brewer's yeast, or a relative, to make a fermented beverage. When Christopher Columbus sailed to the New World, he found the Native Americans already fermenting agave, manioc, and maize. And further north, in your neck of the woods as you like to say in America, they were making alcohol out of birch bark." [li]

I told Gamal about a display I came across as a student on a field trip to a museum along the Oregon Trail and about how the early pioneers on their way west used the succulent growths at the tops of pine trees to brew beer when nothing else was available. There are sugars in pine needles as in other kinds of leaves. They also used what was left of their dried vegetables, purchased before leaving St. Louis, to brew beer if they had any left. They would supplement this brew with fresh blueberries gathered along the trail too.[lii]

"Oh, and I almost forgot," he said, "Beer is a source of B-vitamins, and minerals like magnesium and phosphorus.   Especially the darker beers. Bread and beer are the two most important staples wherever grains are grown." [liii]

AN UNEXPECTED DETOUR

I was curious how Gamal had ended up working with the yeast and finally saw an opportunity to ask.   And why wasn't he still at the university, at least in an official capacity?   He answered as we began exploring a room devoted to Egyptian furniture, which was all wooden.[50]

"My original work was with another species of yeast...one that causes a disease. Thrush. Have you heard of it?"

"Heard of it."

"It is also called candidiasis and if you had children you most certainly would have seen it in the form of diaper rash.   Thrush is a painful yeast infection in the lining of the mouth and tongue, especially for those with compromised immune systems.   AIDS patients for example.   We all have this yeast in our mouths, but our immune systems keep it in place.   Doctors in Japan have found patients with so much *Candida* yeast growing in their intestines, they can make their own alcohol...enough to be legally intoxicated. We all have a tug of war going on inside us between microbes and our immune systems all the time.   Because fungi like *Candida* are more like our own cells than to bacteria, it is correspondingly more difficult to rid them from the body without harming our own cells."

"Trying to kill fungi with antibiotics is about as much fun as trying to teach a cat walk backwards."

"Whose cat can walk backwards?"

"That's okay, Gamal. I didn't mean to interrupt."

He continued explaining about his previous work: "The same antibiotics that work against the fungus damage us, which creates side effects.   I was looking for new and useful compounds, small molecules actually, that could kill yeast blooms that cause thrush while leaving human cells unharmed."

I knew it was true what Gamal said about the yeast being more like our own cells than to any bacterium it shares the soil with, in fact this was the very reason I had gotten the grant in the first place: to make human proteins inside the brewer's yeast...proteins that are more similar to our own than to those

---

[50] The Egyptians also invented folded furniture. A folding chair was found in King Tut's tomb.

made by a bacterium like *E. coli*. A protein being more human means less likely to cause side effects when given as a medication. Gamal had been doing what's called a *drug screen*.

He was silent for a moment, as if trying to make up his mind on whether to let me in on something or not. Finally, he must have decided.

"What happened is that I lost my lab. I tried to hang on…for some time, actually. Even Tahany tried to help out but gradually…what with no money coming in…which of course means no bench space…it is the same for you, isn't it? But at least I was able to turn my attention towards the brewer's yeast. I am doing quite well now thanks to *Saccharomyces*. But if you had told me just three years ago I would be working on making a better tasting beer I probably wouldn't have believed you." I didn't get the feeling he wanted to change the subject so I didn't say anything. "You know, in today's world, your thoughts are still on your career. But if you had lived in ancient Egypt and were in your 30's you would have been thinking about what your tomb should look like. Making arrangements for the afterlife was common even for someone your age in ancient times."

I was only half-listening for I couldn't shake the notion that, due to my losing out on the big anthrax grant, Gamal and I were traveling a more similar road than I had first imagined. At least he was resourceful enough to have landed on his feet…having found a place in industry. A lot of us in the states were turning to private industry too, biotech companies for example, as the government left many young scientists withering on the vine. Even though I was trying hard to stay in academia I was still being forced to switch fields, from a bacterium to the yeast, which may not sound like much but is, in fact, a major step biologically. I'd just as soon *not* have changed fields to be honest. Most microbiologists would definitely consider what I was doing – moving from a prokaryote like bacteria to a eukaryote like the yeast – not to be a very good career move on my part, but I needed to keep the money flowing in. The brewer's yeast was my only chance now. But worrying about finances is hardly new. As Gamal would mention later in the day, even scribes in ancient Egypt had to make extra income sometimes by taking in students and teaching them to read and write hieroglyphs.

Gamal stopped at what was for him a waist-high model of an Egyptian house from the Old Kingdom period, made out of mud bricks; mud being another gift of the Nile.[51] It's where he picked up his lecture on the brewer's yeast.

"If I have given you the impression that all I have told you is proven, I must apologize. In fact, no one knows with any certainty the world's first fermented beverage." He stared off for a moment with a blank expression on his face

---

[51] Dividing up 3,000 years of Egyptian history into the "*3 Kingdoms*" we still use today was the notion of the Greek historian Manetho, who lived during the 3rd century BC.

and I almost regretted asking him about his research. Something was clearly weighing on him and I didn't want to make it worse, whatever it was. I had so many questions, though.

"Did you know my ancestors used to keep bees?"

"I hadn't. There's a lot they don't teach us about your part of the world. Maybe they think we'll learn it all in church."

"They were so skilled at beekeeping, they would use clay pots as hives and transport them up and down the Nile depending on which crop was flowering. It seems honey is a good source of the brewer's yeast."

We moved on with what would become a familiar pattern throughout the rest of the morning and into the afternoon, Gamal talking, I listening, as we wove our way in and out the museum's many rooms. At one point I imagined myself as a thread trailing behind Gamal, the probing needle intent on stitching the museum together like some kind of quilt. I listened and watched because it was becoming clearer just how much he knew about the yeast and besides, you're never supposed to stop a horse from galloping in the right direction just to give him sugar. Still, I couldn't help wondering if the loss of his laboratory hadn't taken some kind of toll on him. Once doing independent research is in your blood...let's just say other things can seem mighty tame by comparison. When not showing visitors around the museum or working with Tahany in her lab, Gamal was probably taking orders from some technician half his age on how much contamination was acceptable in his beer vats; a far cry from being on the front lines fighting a disease.

Another of our unscheduled stops, this one brought about by the soreness in my back, led us to the nearest wall. I leaned against it while Gamal continued describing the first fermented beverages and how they might not have been either beer or wine, but honey. In the Middle East, high caloric drinks like mead materialized, often without any help, discovered after rainwater diluted a honeycomb inside the hollowed-out trunk of a tree. The ancient Egyptians were accomplished beekeepers and, as Gamal pointed out, pure honey is high in sugar and therefore a good source of wild brewer's yeast.[52] It seems the yeast lies dormant, trapped within the viscosity of the honey...that is until the inhibitory sugars get diluted by rainwater, whereupon these same sugars reach a lower concentration and so the yeast springs back to metabolic life, fermenting the sugars into alcohol.

For the Egyptians, honey might have been what for us is the equivalent to freeze-dried yeast on a supermarket shelf. Whenever they needed its power to start a new batch of beer or wine fermenting, they may simply have added a cup or two of honey. Honey would have been what we today refer to as a "starter culture", not so different from how I sometimes transfer *E. coli* from an

---

[52] The Egyptians had over 900 remedies requiring honey as an ingredient.

old Petri dish onto the end of a sterile toothpick to inoculate a fresh batch of broth just by tossing the toothpick in. Sugar is sugar, chemically. The yeast doesn't care where this carbohydrate comes from.

"Yes. It is pretty amazing, isn't it?" Gamal had noticed my increased interest. "My people were the first microbiologists and yet they didn't know about microbes. They would have passed fermentation off as the will of the gods, not so different than the rise and fall of the Nile I suppose...what for them was their recurring miracle. At one time there was no oxygen in Earth's atmosphere and so most cells had to make due by fermenting sugar to get all their energy, like yeast do today.[liv] In some ways, that makes the yeast a relic...more ancient than anything in here, that's for certain."

"Maybe we could start a museum for microbes," I added. "And we could arrange them in different rooms depending on how ancient their metabolisms are."

"It would have to be a big museum, though. Yeast and human cells are separated from a common ancestor by about a billion years." He laughed and I made a mental note to change the subject more often as it seemed to cheer him. I looked around, noticing for the first time that the museum had filled to capacity. Still sore from the previous day, I began looking for a place to sit, except that all the benches were taken by elderly tourists and young mothers caring for small children.

"Ancient Egyptians would have been well aware of changes...of what constitutes a normal fermentation...and felt a certain reassurance whenever it took place," Gamal continued. "Their lives could have depended on it. Ancient people lived in constant fear of famine. They kept careful watch over the changes fermentation brings as closely as they watched the height of the Nile each season, monitored their food's colors and smells, its textures, the roiling of carbon dioxide bubbles as the sugars turned into alcohol and carbon dioxide.[53] If all went well, the food would have changed predictably, from a sweet one to a slightly sour one. Next the distinctive aroma from the alcohol would have begun mingling in the air just above the pot as they whiffed the liquid with their cupped hands." Gamal closed his eyes as he described the scene, "Whenever they saw fermentation taking place, it would have filled their hearts with a warmth I suppose. Fermentation was the first process ever taken from nature and reproduced in a controlled fashion. Here was an event that could be found in the natural world, as unpredictable a place as we can imagine, and yet they made it predictable given enough skill. They improved upon it and eventually made fermentation into an industry. It may have been a source of empowerment and pride for ancient people, allowing them to feel in control of a very mysterious world."

---

[53] If conditions are optimum, sometimes this roiling can be so rapid as to be audible, and is described in ancient texts.

A SHEPHERD'S LIFE

Gamal then explained how the brewer's yeast makes no powerful toxins, yet lives in perhaps the most competitive place on Earth...its soil. It is no coincidence soil was where scientists first obtained antibiotics – penicillin and streptomycin – in the 20th century. There's so much rotting food and competition for it down there, microbes need all the help they can get to survive in places like this, and they have to be good at making use of nutrients because everyone's waging war. But rather than produce a toxin, or an antibiotic, the yeast simply avoids this and instead relies upon fermentation...and patience. Like other fungi, the yeast doesn't have a problem waiting things out. Unlike plants, fungi don't need the sun's rays to make their food.

We walked out into the atrium and then strolled into another musty-smelling room where Gamal picked up his talk. "Strictly speaking, the alcohol is a poison since it is good at keeping unwelcome bacteria out. Unlike most bacteria, the yeast doesn't require oxygen to extract energy from food. It lives at the bottom of compost heaps and so bores it way through rotting vegetation, exploiting and digesting the plant's sugars as it goes, making alcohol while keeping oxygen out. Reducing the competition. There's not much oxygen inside decaying things anyway."[54]

"Fungi are nature's hidden kingdom," I added.

"Yes. And as for bread dough, the yeast was enlisted only later. The very first breads were flat, unleavened breads. The earliest evidence for leavening dough is from 4,000 BC. People in your part of the world would say meat and potatoes. In Egypt we say bread and beer...but they mean the same. The Mexican cuisine is based on not having a fork," Gamal continued, "Flatbreads like the tortilla are an ancient cooked gruel. Wrapping food in flatbread is a convenient way to keep your hands clean. Think of a bean burrito. The very first breads were unleavened and started out mostly as water. Then they cooked and hardened these liquids onto the surfaces of hot stones, producing a bread with very little moisture, one that could be stored and eaten anywhere. It is why we so often think of the nomadic shepherd's life as consisting of unleavened bread. And why Moses' people were in such a hurry once settling to begin leavening dough again. The smell of bread baking in a brick oven or inside a clay pot undoubtedly reminded them of civilization.[iv]

"The world's first comfort food?"

---

[54] Some species of fungus like *Aspergillus flavus* make a poison called aflatoxin so powerful that a single molecule of it can mutate DNA. / Indeed, the most toxic substance on Earth is *not* made by a nuclear reactor, but by a tiny bacteria cell: botulinum toxin (the active ingredient in Botox).

Again, Gamal didn't seem to notice my joke. "Letting bread dough rise is a way to enhance not only the bread's texture but also its flavors...to increase the variety of molecules responsible for taste,[lvi] enhancing its digestibility. Here is where the Jews learned of it. This isn't even the oldest story in recorded history. That distinction goes to the Mesopotamian *Epic of Gilgamesh*," and with that Gamal explained how one of its main characters – Enkidu the wild man – became civilized, made more human if you will, only after drinking beer and eating bread for his first time.[55]

Gamal continued: "Beer and bread have always been closely related, which is why, when we find an ancient brewery, it is always next to a bakery here in Egypt."

"Because both are made of grain?" I ventured.

"And the yeast. These first beers were accidents, intended as liquid breads, but someone left it out too long...long enough for the yeast to settle in from the air, or perhaps from the Nile River's water, to colonize the gruel and begin fermentation."

He explained how the first doughs were made after chewing wheat kernels in the mouth. And how enzymes in our saliva would have begun the process of chopping up the plant starches into simple sugars required by the yeast. It's the reason when you chew a piece of bread or raw potato long enough in your mouth they begin to taste sweet. The enzymes in your saliva are cleaving and liberating smaller sugar fragments from the ends of long starch chains, getting these simpler sugars ready for absorption into the blood.

The first cities came about in Mesopotamia, and increased crowding meant more opportunity for spread of disease, and therefore an increased demand for fermented beverages. Beer and wine have always been safer than water to drink, wine because of its higher alcohol content and beer for its having been boiled during brewing.

"Did you know the Bible never mentions water for drinking? So it's not surprising that half the barley grown in Mesopotamia would have been earmarked for brewing. The first laws in recorded history were chiseled into stone at the behest of King Hammurabi and included regulations on how beer should be sold.[56] He used stone to emphasize the permanence of these laws. Not only was the price of beer fixed by the king in Babylon, but the punishment for owning a beer house and watering down beer for your customers was death by drowning."

---

[55] ~3000 BC Sumerian tablets. Gilgamesh is also the first story in literature to feature a woman character – a Mesopotamian tavern owner who serves Enkidu his first beer.
[56] 1760 BC.

We passed a display of ancient, yet familiar-looking, plowing tools while Gamal added with what sounded like pride that the Mesopotamian bread would have been more like flatbread, not leavened like bread in Egypt was.

"What about wine in Mesopotamia? Didn't they have wine too?"

"Yes, but grapes did not grow as well as grasses like wheat and barely. Grapevines require more attention in difficult climates and therefore would have been tended to by priests in temples. So wine was a luxury, reserved as a holy drink, or used in sacrifices.[lvii] I am going to mention wine and religion more in a moment, but for now I will only call your attention to how similar religious ceremonies are in so many cultures...and to a priest in ancient Mesopotamia tending his grapevines. When you care for a grapevine, you bring it an offering of water, don't you? Then you kneel before it and tenderly sprinkle the water onto the ground...like a libation. Taking care of a grapevine is not so different than kneeling and praying before a god at an altar. In both cases you're making a personal sacrifice. When the Greeks – members of what most assuredly was a wine culture – visited Mesopotamia thousands of years ago, they claimed their wine god Dionysus fled from it as the people only wanted beer."

"But how about the yeast?" I asked, "Didn't they have some idea of the yeast?"

"They would have known of it indirectly. The yeast is why they used straws to sip their beer. Straws helped them avoid chunks of yeast and other impurities floating on the surface. It was all a mystery to them, brewing, leavening and winemaking. It is why there were so many deities for brewing...Ninkasi is the goddess most often mentioned...but there were others...many others...too many to mention here." [lviii]

Gamal continued without taking his eyes off the display, "Personally, I believe Egyptian civilization will turn out to be every bit as old as Mesopotamian, however it is true that the first written evidence for wine and beer come from clay tablets in Iraq. The world's oldest recipe is for brewing beer." [57]

STARTER CULTURE

There is something comforting about history, its timelessness, the way it joins us with the past. All roads lead to Rome as they say and Gamal seemed to enjoy immersing himself in history as much as I did...taking refuge in it almost. And he was smiling a bit more as we ventured further into the museum's interior, passing twin barges with sails unfurled. Gamal was hitting

---

[57] Wine, 2750 BC.

his stride, confidently walking with his hands firmly clasped behind his back, staring intently at the floor, concentrating on his mental notes I supposed.

"Ancient people didn't know about microbes, but like brewers and winemakers today, they did know that the sweeter the starting material, the stronger the beverage would be, which is why the Egyptians and Hittites sometimes used raisins instead of grapes for making wine. There is more sugar to a raisin than to a grape, more starting material for the formation of alcohol. One can get the alcohol content up to 16% by starting with raisins The Hittites were fond of saying that the 'raisin holds the wine in its heart'. It seems wine could be so important that the first thing an invading army did was to destroy all their enemy's vineyards. It was a way of fermenting unrest among the people, of encouraging resentment for their king's lack of authority. If the king couldn't protect his grapevines, then how could he possibly protect his own people? My ancestors also made wine out of dates...in the oasis communities. They carried this date wine in amphorae all the way to the banks of the Nile on the backs of donkeys, where it was offloaded and shipped around the Mediterranean. But do not be fooled. Egypt was always a beer culture. It never became a true wine culture like the Greeks or Romans later on. Still, my ancestors did appreciate variety in their lives, just as people do today."[lix]

Arranged before us inside a long glass case were shelves with toys on them...wooden ones that Egyptian children were given as gifts over a thousand years before Julius Caesar came here, some with the original paint still on them. I tried to picture a young Nefertiti tugging on one as a child.

"They have strings to pull their moving parts. See the crocodile over here?" Gamal indicated by tapping the glass. "Its mouth opens. And this one," he moved to the other side to get a closer look, "It is three dwarves standing in a line. Can you tell? It was carved during the 12th dynasty. Pull the string by way of tiny pulleys and they still dance."

Another statue had been carved into the shape of a person bent at the waist, and it was obvious even after dozens of centuries that it was someone engaged in toil. The figurine turned out to be the main reason we were in the room.

"This person is kneading dough. While the rest of the ancient world was subsisting on flatbread, we Egyptians were taking leavening to new heights. Leavened bread also has more protein because of the yeast. The yeast helped build the pyramids, doing their part by providing nutrition and sanitary drinking water in the form of bread and beer."[58] Inside another glass case

---

[58] The ancient historian/tourist Herodotus compared Greek bread with Egyptian bread, marveling that "We all worry about food fermenting; but the Egyptians deliberately make dough so that it does ferment."

was what appeared to be a chess table, one that was on runners, looking somewhat like my grandmother's rocking chair.

"It is called Senet.   And all Egyptians played it, from the king on down to the workers in the field." [59] Gamal then went on to explain how more variety was obtained by making bread into different shapes, some as crescents, others more pyramidal, or perhaps in the form of a woman.   They also added honey and anise...even cumin.   Bread was so important that they gave up eating it for periods of time during mourning, just as many Christians do today for lent...as a sacrifice.   Bread may also have been toasted and used as stuffing inside fowl caught along the banks of the Nile.   Scenes in Memphis dating back 5,000 years portray every important aspect of baking.[ix]   I was suddenly reminded of the elusive papyrus I had yet to locate.   Maybe I'd try the museum gift shop, or perhaps Luxor, I resolved.

Gamal continued: "They added the yeast by using dough from a previous batch of bread as a starter culture.   They may also have used bread to start beer brewing.   As long as the bread was not heated too thoroughly, the yeast would survive in toast, still able to begin fermentation of the next batch of beer wort given a chance.   Bavaria is a fine example of a beer culture.   Tahany says you have a stop over in Germany on your way back home?"

"I have to change planes.   A layover of about ten hours I think.   Why?"

He apparently didn't hear me as I followed him through the open door and back out into the giant atrium again.

As we walked along, the thought that perhaps Gamal had picked up on the fact that I was one of Tahany's admirers became a more likely explanation for his strange mood.   Maybe that's why he seemed so awkward, unlike the previous night.   For all I knew, maybe she had even mentioned something about my flirting with her.   And he was older than Tahany, while she and I were about the same age.   The thought left me wondering if perhaps Gamal wasn't the quiet, jealous type.   After two days of knowing someone, it's kind of hard to know these things.

"Language is an amazing thing," Gamal began, "From written records, we know that Ramesses III distributed something like seven million loaves of bread to his temples for distribution, and that the average worker got three loaves and two pitchers of beer a day along with some vegetables like onions."   We stood with our backs against the wall as dozens of school children began parading by, their teachers like shepherds, with one bringing up the rear, apparently keeping an eye out for stragglers.

"The temples devoted to Ra, the sun god, had their own vineyards.   We have always stomped grapes in Egypt to make our wine, you can tell by the wall paintings.   They show men holding onto short pieces of rope above their

---

[59] Senet was a favorite pastime of King Tut's. He even had a portable version to take along on his travels.

heads so they can steady themselves while stomping. They knew pressing grapes with stones could break open the seeds, infusing the wine with an unwelcome, bitter flavor. But the wine was not what you would call fizzy today, nothing like Champagne. In fact they went to great lengths to avoid carbon dioxide building up inside their wine. The Egyptians drilled holes in the clay plugs to let this gas escape. Then when the yeast had done its job, they sealed the plugs back up to keep the wine from spoiling due to oxygen in the air."[60]

After the last of the children were past, we found our way into the next room.

"The yeast cell has always been a mystery and would remain so until the Enlightenment in the 1600's, so perhaps it is only fitting that the yeast was so influential in religion. In the Middle Ages its mysterious fermentative powers were referred to as 'godisgood'."

I had wondered about wine and bread's relationship to religion ever since taking communion in church as a child. None of my family could answer my questions. Everyone just performed the ritual. I realized that year that part of being religious meant not asking so many questions.

"Ancient people had many gods and some were caretakers. It is similar to the way you visit a supermarket today. If you want to ward off mosquitoes you buy an insect repellent. When you need to clean something, you purchase a detergent. Well with ancient people they chose a god for what they needed, took it off the shelf often quite literally, and then made arrangements, a kind of a tit-for-tat deal."

"You scratch my back I'll scratch yours I think you mean," I added gently.

"Yes. There is a difference, isn't there? Tit-for-tat implies something bad for something bad, doesn't it? Well for the Sumerians, it was Ninkasi who watched over the brewing, for us Osiris, for the Greeks Dionysus, the Romans of course had Bacchus to call upon.[61] The fear was always a *stuck fermentation* after all that work harvesting and preparing, of where instead of beer and wine one ended up with sour-tasting vinegar or far worse…putrefaction. Vinegar could at least be useful as a weak acid and as an additive to make food taste sour or for preserving drinking water. Homage needed to be paid to the spiritual world, prayers said correctly, wine and beer libations sacrificed onto the ground in appropriate proportions. Yet they also knew how air could derail the fermentation process. One solution ancient

---

[60] Champagne was created by yeast while still fermenting in the bottle. The French were the first to do a *second fermentation* inside the bottle, making the first sparkling wine, or Champagne. The yeast is capable of producing pressures inside a wine bottle exceeding that of a car's tire (4 atmospheres of pressure) which is pretty small compared to *Magnaporthe grisea*, a pathogenic fungus that can produce up to 80 atmospheres of pressure in order to penetrate the tissue of its host plant.

[61] The Romans considered beer a barbarian drink, calling it a poor imitation of wine.

people came across was to use long-necked vessels for storage. This helped keep the air out. Another solution was to fire-harden clay amphorae, making them more...how do you say...immovable? To the air, I mean. Some have even been discovered 8,000 years old."

I provided him with the word *impermeable* and then considered how familiar amphorae had been to me from a youth spent watching the Undersea World of Jacques Cousteau. All ancient shipwrecks on TV seem to have amphorae scattered about them, it's a requirement almost, containers shaped like a person's body, but without a head; just a curving trunk and slender neck. Some had a pointed end so they could be stood upright in the sand and handles at the top so two workers could carry them side-by-side. Different cultures had different styles it seems, yet the same basic shape remained unchanged for thousands of years, right up through the Roman Empire.

"The Greeks believed the god that gave humanity the gift of wine was Dionysus. Did you know his story is similar to the one you have for Jesus? Or that Egyptians were the first to believe in monotheism and resurrection over a thousand years before the Bible was even written?"

"No," I answered. Another thing I hadn't learned in church.

"Yes. Both Jesus and Dionysus performed miracles like turning water into wine, and both of their bodies would come to be represented by bread, their blood by wine...to be consumed during a ritual meal by their followers. Strange how that is, in fact it is probably why Christianity caught on so quickly in Egypt in the first century. And both Dionysus and Jesus rode donkeys and arose from the dead. Both were preoccupied with peace while on Earth. Both were viewed by their followers as someone who could liberate others from a world filled with pain. It is no coincidence the Greeks believed some of their gods were half human, like Jesus. Both came here as saviors, you see? When you stop to think about it, it is not so surprising how Dionysus could also be the god of wine. His father was the most important god of all, Zeus, his mother the mortal Semele. Dionysus ruled dreams and intoxication. When ancient people drank wine they felt more connected to gods like Dionysus I suppose."

In spite of the similarities between the two, I'd barely heard of Dionysus before.[lxi]

Gamal continued, "I sometimes like to picture the yeast as an epic hero, sacrificing itself for the wine as the alcohol gains in strength. The invention of the theater came about as a result of the wine god too. There was one Greek who, in 540 BC got up and acted out the part of Dionysus during a festival. His name was Thespis, and his round stage was inspired by the shape of a threshing floor. Anyway, the debut of Thespis turned out to be the first play in history.[lxii]   I sometimes wonder where Hollywood would be without Dionysus."

"Or the yeast," I added.[62]

Rather than spending our time talking shop, as one might imagine two microbiologists doing, we used the next hour or so to gaze at artifacts without saying much that I remember.   On one level, we were like two schoolboys separated from their group as we took in objects that had been brought here, often at great sacrifice.   Sometimes the temperature inside a tomb in Egypt can be high enough to melt candle wax.   It's one of the reasons the government put electric lighting in Egypt's tombs as soon as it was available. I'm not sure how long we lingered because it's easy to lose track of time in a place like that, perhaps not so surprising Egypt being the closest thing to timelessness I've ever experienced other than sleep.   Hours can slip by disguised as minutes when you're immersed in this degree of antiquity. Trying to keep track of ordinary increments like minutes and seconds seems almost laughable here.

There was a long stretch of silence before Gamal picked up again exactly where he left off by talking about religion and wine, apparently following some sort of an inner script he'd prepared.

"The first person to make wine in the Old Testament was also the first to get drunk on it.   That would be Noah, of course.   Even the practice of keeping new wine in old animal skins is cautioned against in the Old Testament, a reference to the buildup of carbon dioxide in the wine I suppose. The yeast can make this gas so concentrated it literally splits leather apart. The very first miracle performed by Jesus was turning water into wine if I recall.   And as one who was fond of using stories to get his ideas across to people from all walks of life, Jesus likened himself to 'the true vine' while his Father he called the 'winemaker'.   His followers were for him the branches of the vine.[63]   Jesus also referred to himself as 'the bread of life' and his goal was to feed all of God's people.   The Bible mentions wine no less than 200 times.   I am not sure about bread."

In this way Gamal returned again to the topic of bread and the yeast.   "It is still a mystery why God told the Israelites to eat only unleavened bread in the desert.   Perhaps it had something to do with the unclean connotation leavening always carried.   It is in our natures, I suppose, to fear that which we do not understand…things humans have little control over…and it must have been rather disconcerting for ancient people to watch a once flat, inanimate piece of dough rise up all by itself…expand as if it were harboring a

---

[62] It's no coincidence that democracy and acting came about at the same time and place.   In ancient Athens acting would have been a skill necessary to persuade an audience, just as it would have been to sway voters to win an election or prevail in court.

[63] "He who dwells in me, as I dwell in him, bears much fruit." John 15:5

ghost. To them, the flour had become contaminated...possessed is perhaps a better word...possessed by a spirit that could be carried along from the previous batch of dough into the next."

"The yeast made its initial appearance in bread as a spirit?" I wondered. "That's an interesting thought."

"Or perhaps they viewed the whole arrangement more like impregnation," Gamal replied. "I am often reminded of the baker in the ancient city of Herculaneum, the one who before the volcanic eruption had placed carved phalluses above his ovens to ensure his bread would rise and he would be prosperous."

"He was thinking of impregnation?"

"Perhaps. What we today would refer to as inoculation of media. You have seen it in your church, how during communion the bread used to represent Christ's body is always an unleavened flatbread. A wafer perhaps? Pure and round, isn't it? The apostle Paul compared leavening of bread to a sin, and warned that a small amount of fermented dough already raised could do the same to new dough. Too much salt kills leavening, they knew this and today we know why. Too much salt inhibits the yeast's metabolism and so slows the formation of gas. It is why the New Testament instructs Christians to 'have salt in yourselves and be at peace with one another'. And yet it is also says that 'a little leaven leaveneth the whole lump' so I suppose it depends upon which verse you read when it comes to the yeast. [64]

YEAST AS SAGE

"Wine played an important part in bringing about western philosophy too. You have been to a symposium before, haven't you Ben?"

"My first one was in graduate school," I didn't mention the topic was on bioterrorism.

"Well the Greeks invented them. Essentially, they were drinking parties but often with a more profound purpose. The Greeks were the first to democratize wine too. At a typical symposium, Athenians would gather in someone's home, drink wine while enjoying various entertainments...talking about important issues of the day. Sometimes they would discuss subjects like 'what is the meaning of love', or 'what is logic'. [lxiii] People simply showed up uninvited. The only requirement was to be a free citizen. Some who planned on doing a lot of drinking even brought their slaves with them...to carry them home afterwards."

---

[64] Egyptians didn't have soap but used natron salt instead. They also made use of salt as a mouthwash, which would have killed bad breath-causing bacteria.

"History's first designated drivers," I quipped, but Gamal still didn't seem to get my jokes and continued on.

"Plato described a symposium that took place around 385 BC." [65]

"It's hard to imagine the scientific method if not for the Greeks," I added, "Einstein once said that he'd rather study the ancient Greeks than science."

Gamal continued: "According to Euripides, Dionysus gave mankind the simple gift of wine, the gladness of the grape, to rich and poor alike."

"And all this time, no one suspected a living organism was behind the mystery of leavening and brewing?" I asked.

"No one saw the individual yeast if that is what you mean, except as clumps or foam on the surface of the vat.   My ancestors knew that if they used the same vessels for brewing, they would get more reliable batches of beer.   The clay containers with the most cracks and crevices worked best, no doubt because the yeast could hide out, hitching a ride into the next batch. They were chain culturing a microbe.   The yeast co-evolved in the process, maybe at different times and places, but adapt the yeast did, as it is still doing today.   Most yeast float toward the end of brewing.   The term we use in industry is *flocculation*.   There is an interesting quote from Pliny the Elder, the ancient Roman historian, about the Gaul's using this beer foam to make bread lighter, but I cannot remember the exact words, so I will not attempt it."

Gamal talking about the yeast foam took me back to my senior year in high school and our first fieldtrip, the one where we were to observe microbes in nature.   Here I had been expecting to collect samples and take them back to the classroom and look at them under the microscope, but instead the teacher merely pointed out microbes all over the place and we used our five senses to detect them, wherever we walked there were microbes that day. Our senses are enough to see, smell, touch, taste, and perhaps even hear them on occasion.   You can detect microorganisms by whiffing a handful of ordinary dirt.   That day we observed circular orange mats that looked like layers of a giant onion, cut down the middle and alternating between different colors: red, yellow, green, and brown, all one inside the other within the pristine hot springs of Yellowstone.   They are bacteria and algal mats.   The different temperatures allow only certain species to thrive in them.[lxiv]   In the center where the water first percolates out of the ground is the hottest and is where the colors were blue.   Then they transitioned into yellow.   We didn't take a thermometer along because just looking at the different colors gives an idea of the thermal energy the water possesses, which determines the species of microbe living in it.

---

[65] It discusses the meaning of love.   To the Greeks there were different kinds of love (e.g. "Platonic love").   Plato was a student of Socrates, who poetically once said of wine that it could moisten the soul and lull grief to sleep.

But you don't have to travel all the way to Yellowstone to see microbes on any given day. Even the slime on the surface of your teeth you can detect with your tongue if you skip brushing for a day or two is a film of oral bacteria that continually springs back into existence given half a chance. If you have a bottle of wine that goes cloudy then you can blame bacteria...in fact the cloudiness you see are bacteria, millions of them in each drop. You can culture microbes with just a bottle of wine or a bowl of beef broth soup (Leeuwenhoek claimed the best source was in water laced with black pepper). You can even touch microbes with your fingers if you've ever left hotdogs in a freezer too long. That slime is a relative of brewer's yeast, a fungus that thrives in the cold.

On my way to a symposium, I once flew over the Great Salt Lake in Utah at just the right time of day and while looking out the window caught sight of a purplish tint to the waters. This phenomenon was created by another bacterium, one that thrives in the salty waters of the lake, a microbe called halobacter. Even the immense white cliffs of Dover in England are made of the accumulated shells of single-celled algae that lived and died millions of years ago.[66] The black slime that accumulates on shower curtains or the sides of buildings wherever downspouts direct the runoff of rain can be places where bacteria will happily grow. You are experiencing them whenever you wince at the odor of a dumpster on a hot afternoon, and in fact over half the oxygen we breath is manufactured by photosynthetic life in the oceans too small to be seen without a microscope.

Then there are the holes in Swiss cheese, created by gas given off from a microbe called *propionibacterium*, one that normally lives off the protein in our outer skin, and like the yeast, ferments sugar into carbon dioxide. Everyone's seen the orange color on the surfaces of Limburger and Muenster cheese; but fewer know that this pigment is from a *Brevibacterium*, the same species that causes unwelcome foot odor. The microbe consumes protein, whether it's protein in cheese, or someone's dead skin between their toes.[67] It's not picky; both are warm, moist places with plenty of nutrition. Turn over a compost heap sometime and you can feel the heat with just the palms of your hands. What you are detecting is the collective metabolism of billions of unseen bacteria, all furiously burning glucose, not so different than how our own bodies stay at a constant 98.6 degrees F due to our cell's collective metabolisms...the continuous burning of glucose by our ten trillion cells. [lxv]

---

[66] Egypt was under the sea during the Cretaceous period. Over millions of years, single-celled creatures called *nummulites* lived, died, and sank to the bottom. In time their shells built up and were compressed into limestone rock, the oceans retreated and evaporated, and the creature's remains were cut into blocks and used as stones for the Great Pyramid.

[67] The blue color of certain aged cheeses is due to the fungus that makes penicillin: *Penicillin roqueforti.* / Pliny the Elder commented on blue cheese's flavor in the first century AD.

On a molecular level, we're a lot more like microbes than it's comfortable to admit sometimes.

SAY THE MAGIC WORDS

We headed upstairs to the Royal Mummy Room, the soreness in my legs an unwelcome reminder of my overdoing it the previous day. I began to wonder how I was ever going to climb the Great Pyramid at Giza, as I had wanted to since reading Twain's description of it. Some cowboy I'd turned out to be. The only consolation was that my brothers couldn't see me coming up lame halfway up a staircase, one hand on the rail, the other tending to my sore knee. But something else bothered me even more. Tahany and Gamal had been so generous with their time, and yet I hadn't been able to reciprocate much. I made a mental note to find a way to show my appreciation before leaving for Luxor.

The Mummy room was darker and quieter than expected, heightening the suspense the way turning off the lights in a movie theater can do before the show starts, and my eyes took a few minutes to adjust. After a while I began registering details like the condition of the skin and the texture of the dried hair on Ramesses the Great. The ancient Egyptians seem to have been preoccupied with the afterlife and you can tell a lot about a society by what it values. With the Romans it was their baths, a reflection of the Roman work ethic and rich social life, the Greeks had their athletic events, theater, and other arts.[68] While with the Egyptians it seems to have been mostly about pyramids and mummies. If the king were granted eternal life after death, then his immortality would benefit everyone. He became an eternal god. More recently, Egyptologists have come to think of the pyramids as a joint building project that helped bring about not only the pharaohs' tombs but also Egypt itself as a nation. The pyramid builders, priests, and embalmers were the ancient equivalent to NASA's space program in the 1960's. And perhaps not coincidentally, both were concerned with placing someone up among the stars.

"Is it considered disrespectful to display a mummy?" I had to ask since this is the elephant in the room kind of question if you ever find yourself standing next to an Egyptian in a room full of mummies.

"In a way, yes...but in another way, not so much. You see Ben...every time we say one of their names we are keeping them alive. My ancestors believed words had magical, even restorative powers. So in a way, we are

---

[68] The largest buildings in ancient Rome were the public baths, which were "social levelers" because rich & poor alike bathed together and the baths became a symbol of Rome itself. Most towns throughout the empire had at least one public bath.

Secret Life of the Brewer's Yeast

doing them a favor by saying their names...of remembering them after all these centuries."

"If you look closely enough," Gamal continued, "you can see a family resemblance."

I glanced around at the dozen or so pharaohs lying in their climate-controlled cases, which is when I noticed others were also speaking in lowed tones. "It's true," I whispered, "the shape of the nose is recognizable...with a bit of imagination."

Gamal continued: "Mummification was advanced enough by Ramesses the Fifth's time that we can still make out the smallpox scars on his skin, even the remnants of tuberculosis in the spine of this one here." [lxvi]

I was familiar with TB, a disease caused by a bacterium that infects some of the same cells in the human body that anthrax bacteria go after, the macrophages, but with far different results for the patient. Anthrax is a swift disease, TB a more gradual one. In the lab I'd worked with TB, a weakened strain of it, for a time in graduate school before joining the anthrax group. Gamal led me to another king lying in eternal state.

"Meet Seqenre the Second, he was a Theban who met a violent end at the hands of the Hyxos. We can tell by the wounds on his body...five in all. You may count them if you wish. We surmise he was either killed in battle or executed afterwards because the places on his body where he would have been wearing his armor have all been left unharmed." Gamal explained how mummies have been found with hardening of the arteries, even though, surprisingly, they didn't eat much meat. Ramesses the Second may also be known to us as Ramesses the Great, but this pharaoh suffered from arthritis and tooth decay. X-rays of his skull indicate he died of an abscessed tooth. In fact, he's been described as a *dental cripple*. [69]

"Look, he had red hair, but of a darker shade than yours," Gamal said pointing through the glass near his head.

As for me, I was too busy convincing myself I might actually be staring at someone mentioned in the Old Testament, someone who talked face to face with Moses. It seems dental problems were common, given all the sand in their bread. It came from grinding stones used to mill the flour. Eating bread brought with it a hidden price in ancient Egypt. Abrasive sand over time wears away tooth enamel, the hardest substance in the human body, harder than steel, eventually wearing teeth right down to the pulp. [lxvii] I guess they never made the connection. Tooth decay may have been another reason the Egyptians invented the breath mint. I started to feel sorry for him lying there;

---

[69] Current evidence suggests inflammation plays an important role in heart disease. / Within a year after their discovery by Röentgen, x-rays were being used to look inside Egyptian cat mummies in a museum (in 1898).

that is until I thought about his having fathered some 90 children while I hadn't a single one to my credit.

"Didn't the sand lead to lung problems?" I seemed to recall reading something about it as a student.

"It is called pneumoconiosis," Gamal explained. "We see it in mummies all the time."

Oxygen in the air, it seems, combined with moisture is the unseen enemy if you're a mummy. It encourages bacteria and fungi to grow. In the 1970's it was realized that the royal mummies were decaying so steps were finally taken to address the problem.

Gamal somehow managed to find another connection with wine. I had come to realize by now he knew more about the yeast than I could have expected. In fact, I felt as though I could have asked him anything…the true sign of an expert.

Gamal continued: "Ancient people realized air was the enemy and deliberately put additives in wines.[lxviii] Terebinth, a resin from pine trees, was added for its antimicrobial properties. It is one of the traces we can detect with mass spectroscopy, fingerprints we take as an identifier of the ancient wine. There is a wine in Greece still sold called Restina, which has resin added to it. It is an acquired taste, and it turns out I could not acquire it," Gamal made a sour face, "My ancestors knew resin was good for mummification too. It discourages bacteria on the human body for the same reason it does in the wine."

I got my fill of mummies quicker than expected – which is probably why I never wanted to go to medical school – and with my eyes fully adjusted began looking around; wondering what else was in the room.

"Wine, in fact, is the world's oldest medicine still in use," Gamal continued, "Even today we see its health benefits appearing in some epidemiological studies.[lxix] I do not suppose in Montana you have so many health problems?"

I instinctively looked into his eyes to see if he was really serious.

"Well, according to my grandfather, back in the days when Thistle was being founded in the 1860's, they remembered to build a general store, a schoolhouse, even a jail, everything except the cemetery. But since everyone was so healthy, in order to start it they had to go out and shoot somebody."

Gamal managed only the faint traces of a smile. He should have laughed outright because that joke always worked. I leaned up against a railing long enough to give my back muscles a brief respite before we headed into the next room. Try as I might, I just couldn't nail down what might be bothering him. It had little, if anything, to do with my admiring Tahany I had decided by them. It was more like anxiety. He was as nervous as a cat in a rew house. It was as if Gamal was worried about some impending doom hanging over his head. Or maybe it was just his nature. I couldn't tell.

"I LOOKED UPON WONDERFUL THINGS"

"Look," Gamal pointed to an inscription in hieroglyphs on the wall of the next room that turned out to be a replica of a tomb from Saqaara. "As a professor, you will no doubt appreciate this one. It was written by a teacher who lived several thousand years ago and he is saying to his wayward pupil 'I am told you abandoned your studies and whirl around in pleasures, that you go from street to street and the place stinks of beer every time you leave.' And here is another. It is by a girl who describes a boy in the following way. 'He is like a date cake dipped in beer'.[70] I imagine she was in love. Egyptian women had more rights than other women in the ancient world. Did you know? We had taverns where men and women could socialize together. We know there were some women doctors in ancient Egypt but we do not know how many." [71]

Gamal noticed me looking around for a bench and so offered to sit while he continued. I was thinking about the Citadel from the day before, concluding that was where I must have done most of the damage, which is what I get for trying to impress a pretty Belgium tourist. A 50-year-old man with sad eyes and a bad case of nerves was outlasting me in a museum, and a pretty slender man at that.

"While we sit I will explain more about medicine before we go back down the stairs. The alcohol in beer and wine is high enough that it can dissolve some substances ordinary water cannot...valuable substances with intrinsic medicinal value. Water is an excellent solvent, but with alcohol, it is even better at dissolving organic things. So they used wine and beer as a base to dissolve drugs from plants you see?[lxx] The yeast with its ability to make alcohol out of sugar juice...it...I am not quite sure how you pronounce it in English...but it keeps the medicines from forming the hard bit," [72] he held up his thumb and forefinger as if holding a small, invisible pebble.

"A precipitate?" I ventured.

"Yes. Thank you. A precipitate."

---

[70] When a man asked a woman to marry him in ancient Egypt, he may have done so by offering her a sip of his beer. Egyptians probably married for love.
[71] Egypt had the first female monarch in recorded history.
[72] Egyptian papyri list over 100 medicines requiring beer as an ingredient (1600 BC). / The first physicians to specialize in areas of the body like the eyes, teeth, and the gastrointestinal tract, were in ancient Egypt.

We sat while watching the other tourists pass by.   It can be strangely comforting to see people from your own world with electronic devices like cell phones and iPods held up to their ears pulling you back to the present.   It's easy to feel out of sorts, lost to a time warp that a museum can create.   It can be a bit overwhelming after awhile. Gamal, however, was still thinking about wine's role as a medicine.

He continued: "The Romans had their laxative wines and as late as the 1800's many wines were still being used by doctors in the West to treat various ailments.   When I was at Cambridge late one evening I came across an old book in the library written by a physician prescribing various wines, a port wine for women in the midst of childbirth or another useful as an antiseptic to dress wounds, a sherry for someone suffering typhoid fever might also be in order.   When Alexander the Great brought his army through Egypt and then on to India, he added wine to their water to ward off disease. He founded Alexandria and as a Macedonian he would have preferred drinking his wine undiluted, something the Greeks in Athens would have been shocked at."

"It's one of the few things I remember from Ancient History 101," I quipped, but still no reaction from Gamal, perhaps the only Egyptian I'd seen so far with an aversion to smiling.

Gamal continued: "His army even had a mobile city that followed them around, bakers baking bread for example.   After Alexander led the way to Egypt, the Romans found their way here too."

We headed back down towards the Greco-Roman exhibit.   Egypt was a culture that maintained its independence even after invasion by the Hyksos and later the Assyrians, the Persians, and lastly Alexander.   In some ways Egypt was never conquered, at least not until the end.   Some claim Egypt changed her invaders more than the other way around.   But with Rome things were different.   The Romans were the only occupiers who turned Egypt into a vassal kingdom.   The land the pharaohs once ruled over became Rome's breadbasket.   To their credit the Romans did adopt Egyptian ways, and yet they never left a significant monument of their own here.   Still, they did colonize her, and some of the most beautiful, well-preserved mummy coffins in Egypt were commissioned by the Romans.   They're made of wood and the owner had his or her face painted on the lid.   It's striking to see an intricate portrait of someone who lived two thousand years ago; exactly how they would have looked in life displayed over top of their mummified remains.

Gamal continued: "When the Greeks came to Egypt, they couldn't help noticing how only priests and the wealthy drank wine.   This did not sit well on the Greek mind.   Here was a people who gave wine even to their slaves because they believed it imparted strength and vitality.   Even Greek children had their own tiny vessels specially for drinking wine."

"The Greeks did democratize wine then," I asked, echoing what Gamal had said earlier, just to make sure I'd heard him correctly.

"Yes, but the Romans would take this further. Everyone including the children drank wine in ancient Rome. The Romans had a saying – In Vino Veritas – in wine there is truth, because they knew of wine's ability to loosen the tongue…to act as a social lubricant. The Roman historian Justin declared winemaking to be one of the three greatest achievements of civilized people, the other two being a constitutional government and urban life."

We came upon a set of glass cups lying on their sides with pointed ends. Noticing my interest, Gamal explained how the pointed end was so the drinker couldn't set his glass down until all the contents had been emptied in a single gulp. Ingenious, and a typically Roman solution to a problem.

Gamal explained how the Romans were the first to experiment with glass wine bottles but are mainly known today for using wooden barrels. Apparently they learned coopering from the Gauls.[73] Wooden barrels were not only more airtight than clay, but could hold more wine and, unlike amphorae, moved by a single person rolling it on the ground. Sometimes they aged wine for ten, twenty, even one hundred years or more if destined for the emperor's table. In fact, the Romans aged their wines so long that red wines could lose their color and take on an amber tint.[lxxi]

It is tempting to believe that things were always purer in the past. But, as Gamal pointed out, the Romans put what we'd recognize today as additives in their wines, a surprising variety in fact. They diluted wine with seawater by as much as 50 percent and having a sweet tooth meant the Romans added honey whenever possible too, sometimes as much honey as the wine itself.[74] They even added ground up seashells, flour, herbs, and marble dust as these ingredients would have cut down on the acidity caused by organic acids, which create a sour note, allowing the sweetness of the grape's leftover sugars to predominate. There was still no pure sugar in ancient times so they made due with whatever they had.

Another thing they might do was to boil the fruit juices, "reducing them" as a chef would say today, and then add this concentrated sap to the wine. They did this boiling in lead containers. In fact, the Roman orator Cato preferred lead boilers because he said it imparted a sweetness to the wine. The natural acidity of the fruit combined with heat was enough to liberate lead from the boiler out into the wine. Chemical analysis of a victim at Herculaneum indicates he suffered lead poisoning, perhaps from sweetening his wine this way.[lxxii]

Gamal took off his shoulder bag and laid it on the floor, then pulled out a hefty stack of papers, explaining: "We can learn a thing or two from Roman

---

[73] The Gauls sometimes traded a slave for an amphora of Greek wine.
[74] The different varieties of honey would have provided more choices of wine.

burials." After shuffling through the stack, he came across what he was looking for and began reading.

"The love of wine was even expressed by the deceased. On one Roman tombstone a citizen had carved the following epitaph: 'In the ground I lie, who was once known as Primus. I lived on Lucerine oysters, and often drank Flarenian wine. The pleasures of bathing, wine, and love aged with me over the years.' Here is another, also from a tombstone. 'Flavius Agricola was my name...friends who read this listen to my advice...mix wine, tie the garlands around your head, drink deep. And do not deny the pretty girls the sweets of love'."[75]

With the unsettling image of a lovesick toga-wearing Flavius in my mind, I continued walking behind Gamal, then my mind drifted onto another subject, this time Mark Twain and some of the experiences he wrote about on his trip to the Old World. One of their stopovers was the Roman town of Pompeii, in southern Italy and he had been surprised by the amount of excavation already done there. Pompeii had lain undisturbed, buried by a volcano for nearly two thousand years. It was starting to get late and the crowds were thinning as I noticed Gamal leading me out of the Greco-Roman room and back into the atrium.

"The Romans even added an incense called myrrh to their wines. Can you guess why?"

"Taste?" I suggested.

"Perhaps," Gamal answered courteously. "But more likely it was because of its pain-killing properties. Myrrh was an 'aspirin' if you like, and would have made inferior wines taste better...perhaps by dulling the senses of the drinker."

The gift shop didn't have any aspirin or papyrus souvenirs for that matter and the soreness in my back and legs returned full force. At least it was easy enough to locate an unoccupied bench so we sat and talked before going back upstairs to the Tut exhibit. Missing the last two rooms downstairs would be my punishment for overdoing it the previous day, and for eating so many doughnuts during final exams week.

Gamal continued: "Pliny the Elder recorded how the Gauls used foam from their beers to make bread dough rise. In effect, they taught the Romans to recycle yeast, using leftover *beer yeast* for leavening their bread dough. Bakers have always had a long relationship with brewers. Bakers got their yeast from breweries as late as the 1800's...that is until they started

---

[75] Our use of tombstones goes back to ancient Egypt; indeed the oldest stone building in the world is in Saqqara. When an Egyptian wanted to say someone had died, they explained that he "went west". / Among ancient Roman cities in Italy, Pompeii seems to have had especially strong ties to Egypt. Scenes on the walls of homes there often include hippopotami and pygmies from along the Nile.

cultivating their own, more bread-friendly strains.[76] The same way dogs can be selected into different breeds...the yeast was slowly undergoing its own conversion...genetically becoming more like the *baker's yeast* we know today. Most vineyards and breweries these days use special strains of yeast so they can control the quality of their finished product more carefully and reproducibly."

We sat for several minutes more without speaking. Then Gamal came up with an idea.

"I have a thought. Since you have a layover in Germany, perhaps you wouldn't mind taking a side trip for the day...down to Munich. They have wonderful trains in Germany and Munich is a fine example of a European city that emerged out of the Middle Ages because of brewing. I have a colleague there who might be able to help you. He is retired. And I should warn, a bit eccentric. But he might know a thing or two about your project."

"Yes. I'd like that, Gamal. I wasn't relishing the idea of spending ten hours in an airport bar anyway."

"Very good. I'll contact him then. And there is just one more thing," Gamal reached into his satchel, pulling out the same stack of papers he'd read from before, then lowered it between the both of us on the bench, "Tahany says you have a cousin in New York? And she is in the publishing business?"

I vaguely recalled having mentioned something about it in one of our e-mails.

"Yes," I answered hesitatingly, "but I only see her at Christmas...or whenever she decides to come home for a visit."

"Well, I was wondering if...if it wouldn't be too much of an imposition...you see I have been working on this for a very long time now." Gamal put his hand on top of the papers. "It is my manuscript...and has taken me all of four years to complete, but it is finished, at least I am hoping so. It is all about the brewer's yeast, Ben. I hope it is not too much of an inconvenience? If it is, please say so. You are the first person I have shown it to. Except for Tahany of course. It concerns the brewer's yeast and its place in industry and in history. I was wondering if perhaps you could find someone who might want to publish it. I would be most grateful. Maybe you could look it over too...for spelling errors and the like...if you can find the time," his hand was actually trembling. The stack of papers obviously meant a great deal to him.

"I'd be happy to, Gamal," I said, barely able to contain my surprise, yet also relieved that I now had a way to repay he and Tahany for their time. "I've never known anyone who's written a book before."

---

[76] By 1887 the Pabst brewery was adding pure strains of yeast to brew their beer.

"Well it's not a book...not yet anyway," Gamal carefully guided the manuscript back inside its Manila envelope and then handed it over to me. I tucked it under my arm as we went upstairs to the King Tut exhibit, which was everything the guidebook promised and more if that's possible, but oddly enough we didn't talk very much after that. It was as if all that needed to had already been said, the dead having been disturbed enough for one day. So most of what was left of the visit was taken in silence, paying our respects to this civilization that had lasted so much longer than any other but, in the end like a candle's flame, it too flickered out.

Gamal and I were from different cultures and yet we were more alike than I had first allowed. Maybe I shouldn't have been so surprised, though. I often see bits of myself mirrored back at me in other scientists from time to time. It's kind of reassuring and yet a bit unsettling because it makes one wonder if there's enough room for someone like yourself sometimes. But that probably says more about me than anything.

As the turnstiles came into view, Gamal turned to me, looked up and asked, "So is there anything I did not get to?"

"Yes. There is one thing that's been on my mind since I got here. Why does everyone in Egypt smile so much?" I was collecting opinions on this matter, having already interrogated Tahany's brother and just about every other Egyptian I'd been able to strike up a conversation with.

"I don't know. Do we? I suppose when you live here all your life you are like the perch swimming in the Nile. The last thing you are in a position to comment on is the water. Or as Tahany says, the camel never sees his own hump."

I shrugged, which must have prompted Gamal to come up with some kind of a theory.

"Perhaps friendliness says something about desert communities in general. Perhaps you really miss the companionship of others when you do not have it for long periods out in the desert."

We shook hands and as we parted I felt the sense of relief that comes with the realization that you didn't do anything to offend someone like you thought you might have. Maybe I'd feel that strongly about something I'd write too someday.

I vowed to get back in touch with him when I returned from Luxor to finalize the meeting with the microbiologist in Munich. After reading the manuscript I would come to realize just how much he and Gamal had communicated. In fact, most of the quotes seem to have been attributed to this Wolfgang character. If I learned half as much from the self-proclaimed "Yeast Master of Bavaria" as I did from Gamal, I was definitely looking forward to our meeting however brief it was to be.

SECOND WIND

That night was my last in the Bedouin before beginning the second leg of the journey down to Luxor.   The anticipation was welling up again just as it had the morning I left for Egypt; strange the way excitement interspersed with overwhelming fatigue can come and go on a long trip, kind of like the tide coming in and out.   Except this time I was journeying farther into the heart of Egypt, several hundred miles up the longest river in the world, which would allow me the opportunity to compare different cities in Egypt for the first time. The guidebook provided some hints about what to expect, as well as displaying some nice snapshots, the kind of photos I'd like to take more often if I had the time.   On the east side of the Nile it promised Karnak temple, the largest religious complex ever built.   In fact, its map took up two full pages. The book called it "the most expansive open-air museum in the world".   Just to the west lay the Valley of the Kings, where tombs had been carved into hillsides thousands of years ago and could be toured by the still-living, among them the final resting place of Tutankhamen himself.

But my trip wasn't just about Egypt, or Tut, but also the brewer's yeast and how Egypt came to appreciate its special gifts and how both changed one another, and as if our conversation had ended too abruptly back at the museum, I felt the urge to pick Gamal's manuscript back up and thumb through it one more time before turning off the lights and resting my eyes.

I stopped midway through his chapter on Rome to ruminate on what I had just read.   It seems Greek bakers taken as slaves taught the Romans to leaven bread in 168 BC.   Like the Egyptians, the Romans leavened bread by inoculating new dough with old dough from a previous batch.   Bread apparently caught on so well that by the Emperor Augustus's reign there were more than 300 bakeries in Rome alone.   One Roman baker even had his tomb built in the shape of an oven.   Leavened bread found its way into religion, too, in fact the Romans had a festival devoted to bread on their calendar, complete with decorated donkeys paraded through the streets sporting baked rolls fashioned into jewelry and worn around their necks.

I thought back to freshman biochemistry and the way the same simple chemical reaction that takes place inside bread dough is also what makes beer and wine so valuable.   With bread, it's the carbon dioxide one wants, for this creates the tiny air pockets that can, thanks to the heat of the oven, expand and force the flour apart, adding to the unique texture and mouthfeel of fresh-baked bread we all crave, while with beer and wine it's the other product of fermentation, the alcohol, that matters most.   The alcohol is present in the dough too, but only for short while.   Because of its low boiling point the alcohol evaporates away during baking, contributing to the complex aroma of fresh bread. During leavening, the brewer's yeast is consuming the plant sugars and shredding them into carbon dioxide and alcohol, whether from dough or grape juice, it doesn't really matter to the yeast what the source is. This extraction of the chemical energy locked within the bonds of a sugar molecule is how most cells on Earth get their energy to manufacture ATP molecules too, ours included.

I kept reading longer than expected for it seemed Gamal's infectious passion for the yeast had rubbed off on me. According to Gamal, from the middle of the second century BC on, bread was within easy reach of most Romans. Even spectators in the Colosseum watching gladiatorial games and wild animal fights could enjoy bread during a match. They purchased it from stalls, baked in special molds, perhaps even engraved with the image of a gladiator who was in a fight that particular day. The baker may have added his own name to the loaf too...for advertising. And in addition to staging gladiatorial games for free, Roman emperors also gave out flour and bread to Rome's less fortunate, hence the phrase keeping the masses happy with "Bread and Circuses."[lxxiii]

It's strange how nature herself – rather than humans – has been what best-preserved ancient cities for centuries. Angkor Wat and Machu Pichu were abandoned and forgotten about while being swallowed up by jungle. Luxor temple, where I was headed the next day, was buried by sand and even had a village on top as late as the 1800's, when it was moved and the site finally excavated.

In AD 79, a powerful volcanic eruption and the ensuing lava and mud buried two cities in southern Italy, Herculaneum and Pompeii, effectively allowing us a glimpse into what life was like 50 or so years after Jesus walked the Earth. Of the two, Herculaneum was more completely buried, and in fact two thirds of this town still remain underneath the modern city of Ercolano. Naples and Rome on the other hand, have been continuously inhabited and therefore, like portions of the Roman Collosseum or the great pyramids at Giza, were slowly built over top of, or carted off and destroyed.

Pompeii it seems would have been a difficult place to go around thirsty in ancient times. Excavators have uncovered no less than 200 places a Pompeian could have purchased wine, including bars, taverns, brothels, and snack stalls, which were a first century version of a fast-food restaurant, in this town that once boasted 20,000 inhabitants.[lxxiv] According to Gamal, during important gladiatorial matches, vendors would set up temporary stalls outside the Amphitheater to dispense wine to a thirsty crowd.[77] On just one block alone near the public baths of Pompeii there were eight bars selling wines,

---

[77] Roman trackers near the Black Sea captured lions for gladiatorial games by getting them drunk on wine first. They poured the wine into a trough near the animal's favorite drinking hole.

including one local vintage called Vesuvinum. Pliny the Elder, who would also perish in the eruption of AD 79, had categorized 91 different wines available in the Roman world. [lxxv]

Back at the museum I had felt some pride when Gamal entrusted me with something that obviously meant a great deal to him, so I gently slid the manuscript back into its envelope, closed the metal clasp and stowed it away inside my bag before turning off the lights and trying to get some sleep. Tomorrow would be a long day of travel and there was yet another reason the bag would never leave my sight. But I still couldn't get to sleep until I turned the lights back on and dug out my other book, the one that meant so much to a 13-year-old kid growing up in Thistle.

I remembered how Twain had described Pompeii and for some reason wanted to read from it one more time, so I flipped through the pages with the worn-down corners until I found it again. The *Quaker City* had anchored off the coast of Italy in 1867 while Twain, along with some shipboard companions, ascended Mt. Vesuvius riding on the backs of mules. Pompeii had been rediscovered not long before and Twain was surprised at how extensive the excavation had been.[78] Walking around Pompeii's abandoned streets the bachelor on the verge of becoming America's most important writer was moved by how devoid the place seemed, yet still recognizable after nearly two millennia. I savored his words one more time. Twain wrote:

> "...I went through shop after shop and store after store, far down the long street of merchants, and called for the wares of Rome and the East, but the tradesmen were gone, the marts were silent, and nothing was left but the broken jars all set in cement of cinders and ashes; the wine and the oil that once had filled them were gone with their owners."

Exploring still further Twain came across one of Pompeii's bakeries. He wrote:[79]

> "In a bakeshop was a mill for grinding the grain and the furnaces for baking the bread; and they say that here, in the same furnaces, the exhumers of Pompeii found nice, well-baked loaves which the baker had not found time to remove

---

[78] Some roads in Pompeii had crushed wine amphorae used as pavement called *opus signinum*.
[79] Pompeii's bakeries were on 4 main streets, yet the town had no street signs. Residents of Pompeii would have navigated around the city by using landmarks or drawings on the walls of buildings depicting various trades that took place there.

from the ovens the last time he left his shop, because circumstances compelled him to leave in such a hurry." [80]

I turned off the lights and fell into a sleep so deep I don't even recall dreaming. With no particular time to get up I was barely aware of the call to prayer going on outside and was rewarded with the first truly good rest I'd had since arriving in Egypt.

Feeling refreshed, I finished my complementary breakfast while glancing at the pictures in the newspaper Tahany's brother had left outside my door, loaded my backpack and then left the key inside the room, only to come across a handwritten sign posted on the elevator door that read "Elevator in Despair." Now that's interesting, I thought. Only in Egypt do the people go around smiling while the elevators are sad. Somewhere on my way down the stairs it occurred to me that whoever wrote the sign probably meant to write the word "Disrepair". Still hungry, I checked out of the Bedouin and spooned some fuul sprinkled with olive oil inside an empty piece of flatbread, confident enough now in my abilities to handle traffic in modern Cairo, so much so that without consulting the guidebook I caught a public bus down to the Giza plateau. I was making progress; maybe not so over the hill as I'd begun to fear. Maybe I still had some scraps of youth left in me yet.

Planning was essential to the builders of the pyramids and work was done while the Nile was at its height. It was easier to float the giant stones to the site while the fields lay inundated and the construction project also gave the farmers something to do. A pyramid worker was fed well and got the best medical attention when needed, which would likely have been often.[81] The Pharaoh had a special city built for his workers, one that had a fish-processing plant, some bakeries, a beer factory and even its own cemetery.[lxxvi] Experts today believe it was the building of the great pyramids during the Old Kingdom that helped provide the first sense of social cohesion for Egypt.[82] Giza would have been home to a diverse assortment of characters from all over the kingdom, both upper and lower. When not constructing their tombs, Egyptians shared ideas, styles, even taught one another songs, games, and how to dress. Another bread vendor crossed in front of me with a wide tray balanced atop his head, apparently unaware of the contrast he made with the modern world all around. Even today the Egyptian government subsidizes bread. In fact it was dirt-cheap. The guidebook claimed I could have found over 80 different varieties in Cairo if I'd taken the time to look. [lxxvii]

---

[80] There were at least 30 bakeries in Pompeii and the one Twain visited was discovered with 81 round, leavened, and charred loaves still inside its oven.
[81] Now believed to be among about 10,000 to 20,000 other workers.
[82] Sometimes the Old Kingdom is referred to as the *Age of Pyramids*.

When Twain visited the pyramids, he was hustled to the very top, or "dragged" as he put it, but nowadays there's no climbing all the way up. In fact, I had gotten there so late in the day that the 300 tickets giving tourists access to Kufu's burial chamber on the inside had already been distributed. The entrance I would have walked through had I gotten there sooner was created by an 8th century caliph mentioned in *1001 Arabian Nights*, the same one who bashed a hole into the side mistakenly believing the pyramid would be filled with gold.

Fortunately, standing outside and looking up turned out to be enough for a first-timer and I made a mental note to get here earlier on the way back from Luxor. I watched in fascination as another tour bus pulled up and the postcard sellers closed in for the attack. When Twain's party passed by they too complained of being hassled for baksheesh.[83] Even in the 1840's, tourists complained of graffiti they found at the tops of the pyramids. I closed my eyes, trying to imagine Cleopatra giving Mark Anthony a personal tour here two thousand years ago in her chariot. Strange to think how she wouldn't even have known whom King Tut was, let alone have been able to locate his tomb. That's how good Tut's successors were at making sure his name all but disappeared from Egyptian history, as if he had never even existed.

When Twain was here, people were just coming around to believing microbes could cause infections. In the 1850's, during the height of the Crimean War, ten times more soldiers would die from unseen bacteria in drinking water than from bullets and bayonets of the enemy.[lxxviii] Most, including a young nurse who ventured to the battlefront from England to care for the wounded, thought these diseases were spread by "bad air", the so-called *Miasma theory* of disease that was then popular. But in time this same nurse became convinced Miasma was wrong. At her insistence, ventilation was improved and sewers installed in hospitals and the death rates dropped from 42% down to just 2%.[84]

My guidebook had a list of some of the well-known tourists throughout history that stood at the foot of the pyramids where I was now. [lxxix] One of them was that same nurse, the Lady with the Lamp, Florence Nightingale, which was what got me to thinking about her. She wrote that Egypt's pyramids seemed as if they were boring holes in the sky, while the writer Gustave Flaubert claimed they were trying to come down and crush him. Always the opportunist, Napoleon saw the pyramids as a way to increase his chances for victory by reminding his soldiers just before doing battle with the

---

[83] There were so many opportunities for horse and camel rides around the pyramids in the 1860's that some tourists complained Giza "smelled like a stable".

[84] She later published her methods in a book in 1860 called *Notes on Nursing*. The US government during the Civil War sought out her advice on improving sanitation in field hospitals.

Mamluks that 40 centuries of history were watching over them. And there's little doubt his men would have been impressed. The Great Pyramid was over twice as tall as any building they would have seen during their lifetimes. In fact, the Great Pyramid still weighs more than 17 Empire State Buildings combined in spite of Saladin's requisitioning efforts.

Wondering how they were built has been a question tourists have asked for thousands of years and I turned out to be like the others. It's been calculated that workers managed to lay a two-ton rock on average every three minutes...for 23 straight years.[85] These weren't slaves as had been assumed since antiquity, either. As for me, I came to believe they did it mostly out of pride. The guidebook claimed the average lifespan of a pyramid worker was 32 years and like so many who've visited, I too became reflective, thinking not only about the pyramids, but also of time and biology and how short life can be and how diverse it is too, and in so many ways, more than a lifetime's worth of study would ever allow any biologist to uncover.

There are bristlecone pine trees still growing in California that were mere seedlings when the pharaoh and his scribes set about directing the layout of the Great Pyramid, while at the other extreme are insects I swat at all summer long that will live out their entire lifetimes in less than a week even if I don't manage to get one of them.[lxxx] Bacteria like anthrax – which simply divides n two over and over – can be thought of as immortal, while there are other bacteria that may reproduce only once every century, if that. It was the first time I'd thought about the anthrax grant since the previous day.

From this vantage point, I looked out as far as I could into the distance and caught sight of a train winding along the Nile, appearing almost like a toy. It was mingling with the dusty air and headed in the same direction I would be going later in the evening. Another wave of excitement rolled through me and I had the sudden urge to begin on my trip down to Luxor then and there, but instead settled for walking around the pyramids and chancing the Sphinx's gaze one more time.

Gamal had mentioned how wheat, beer, and bread were so important to Egypt's stability that the pharaohs ordered their soldiers to help out in the fields during harvest time. [86] Egypt was the granary for the ancient world, in fact Egypt's wheat supplies provided a third of Rome's needs and helped keep Constantinople Christian for centuries after the Roman Empire fell in the west.[lxxxi] Gamal said something else I'd been thinking about, too, about how

---

[85] Even the smaller stones weighed 2 tons and workers were sometimes divided into teams that competed with one another during construction. / Napoleon brought a mathematician to Egypt who calculated that there were enough stones in the pyramids to build a wall between Germany and France 3 meters (10 feet) high and 1 meter thick.

[86] Egyptian soldiers were paid 10 loaves of bread a day and carried flour to make bread inside mud ovens they'd build on campaigns.

it wasn't unusual for boys to marry at 13 in ancient Egypt and here I was pushing 40 still without a wife.

After my walk I had a few more hours to kill, recalling that I'd placed the copy of *Innocents Abroad* on top of my clothes in case I wanted to read from it. I carefully thumbed through the worn, tissue paper-like pages until I came across Twain's description of the pyramids as he found them in 1867. He wrote:

"At the distance of a few miles the pyramids rising above the palms looked very clean-cut, very grand and imposing, and very soft and filmy as well. They swam in a rich haze that took from them all the suggestions of unfeeling stone and made them seem only the airy nothings of a dream – structures which might blossom into tiers of vague arches or ornate colonnades maybe, and change and change again into all graceful forms of architecture, while we looked, and then melt deliciously away and blend with the tremendous atmosphere...A laborious walk in the flaming sun brought us to the foot of the great pyramid of Cheops. It was a fairy vision no longer. It was a corrugated, unsightly mountain of stone....Each of its monstrous sides was a wide stair way which rose upward, step above step, narrowing as it went, till it tapered to a point far aloft in the air. Insect men and women – pilgrims from the Quaker City – were creeping about its dizzy perches, and one little black swarm were waving postage stamps from the airy summit – handkerchiefs will be understood."

I looked up and tried to imagine how small a person would seem from the very top. Twain climbed to the summit after paying some "draggers" to take him there. He wrote of it:

"Each step being full as high as a dinner table; there being very, very many of the steps; an Arab having hold of each of our arms and springing upward from step to step and snatching us with them, forcing us to lift our feet as high as our breasts every time, and do it rapidly and keep it up till we were ready to faint...who shall say it is not lively, exhilarating, lacerating, muscle-straining, bone-wrenching, and perfectly excruciating and exhausting pastime climbing the pyramids?...I iterated, reiterated, even swore to them that I did not wish to beat anybody to the top; did all I could to convince them that if I got there the last of all, I would feel blessed above men and grateful to them forever."

It's just as well tourists can't get to the top anymore these days, I concluded. Twain had already done a better job of describing it than I could have.

I closed the book that helped bring me here, put my glasses back on, and looked for another train but saw only the spray from irrigation sprinklers and a flock of birds providing any movement. I marked off another experience from my list and was about to begin a new one. The ancient capital of Thebes (modern day Luxor) lay ahead to the south. Every culture has its creation stories and it's interesting, I thought, how the ancient Egyptians believed that in the beginning there was only water…that is until the Nile receded, providing everyone with land.

I got myself to Ramesses Train Station too early and found I had a couple more hours to kill after locating my upper bunk. While watching the porters making up the beds and with what was left of the sun filtering in through the coach's window, I read some more of Gamal's manuscript, drawn to the past, but was then seized by anticipation of what lay ahead and replaced it with the guidebook.

According to a sidebar, the former Baptist minister and temperance advocate Thomas Cook didn't begin bringing tourists all the way to Luxor until 1869,[lxxxii] but Nile cruises quickly caught on; becoming popular with Europeans and even Americans by the late 1800's. A young Theodore Roosevelt would become a lifelong hunting enthusiast after shooting his first bird from the deck of a houseboat on the Nile while traveling with his family upriver in 1872.

My stay in Upper Egypt was relaxing overall and one of my most enduring memories will probably always be of sitting on the balcony of my guesthouse watching the sun set over the river each evening, the tall shadows thrown down by the ancient arches of Karnak Temple seemingly intent on swallowing everything within reach. One of the best things about Luxor was being able to stay right within the ruins. There were times I thought of Tahany, which heightened my loneliness, but at least I had the consolation of knowing she had found someone good for her. Science has its share of couples. When you do research, you usually spend so much time in a confined lab it's not that surprising relationships develop.

Luxor was more tourist-oriented than I expected but less humid[87] and the backpacker guesthouses and Internet cafes stood juxtaposed within the ruins.

---

[87] Less rain in Upper Egypt is why – during rare downpours – tomb raiders were able to locate ancient tombs. They knew where to dig by noticing where the water disappeared down into the ground.

On my first day I rented a rickety bicycle and took it with me on the ferry across the river to the west side. Waking up in the east and traveling west could have been a daily ritual for an Egyptian artisan, a stonemason or a painter perhaps, carrying the tools he paid for himself to work each day. He likely lived during the New Kingdom period, when Thebes was rising to prominence as a religious capital and he no doubt worked for the king as a member of a team. His pay would have been in beer and food, and if he was exceptional, perhaps some precious metals could be counted on as well. If he were young and talented but not very experienced, he would have been put to work on the less important tombs, only later having been allowed to work on the royal ones.

I used my book's insert map while peddling and sweating my way around the west bank throughout the Valley of the Kings in increasing increments each day. King Tut's tomb was along the route and turned out to be smaller than I expected. His mummy has recently gone on display there, too. I also learned he'd been married at 12, not 13 (my first possible correction for Gamal).[88]

The second night back at the guesthouse I became aware of a change taking place. For one thing, I felt only the vague remnants of where the sharp muscle pains had been in Cairo, replaced now by a dull ache, a good kind of pain – if there is such a thing as good pain – and I even caught sight of myself one evening in a full length mirror, noticing that my pants were fitting better, which made me look like I was standing more upright. I was becoming a well-oiled machine…yet another unexpected gift of the Nile. I tried to make some headway on Gamal's manuscript but most nights I wasn't able to concentrate enough to give it the attention it deserved. It felt almost sacrilegious coming all the way to Thebes and then not learning more about Thebes, so I read all I could in Luxor while peddling around town – tourist pamphlets, local guidebooks left lying around the guesthouse, anything about the area I could find from travel agents and museums, even pamphlets describing the local geology and soil put out by the government.

Several days passed this way, during which time I had built up my own library of material and a comfortable routine, exploring what I hadn't seen of the ruins the previous day by bicycle, then returning to the guesthouse with just enough time to watch the sun set over the river and enjoy a cold bottle of beer in front of me. Then, for some reason I suddenly up and decided to return to Cairo early, to rediscover its fuul stands and other surprises, its colorful hats and scarves, its children playing jump rope in the cemeteries, the sense of cohesion that can make a place special and that Luxor seemed to

---

[88] Tut's wife, also his half-sister, was just 13.

lack somehow.   I missed the human activity apparently, which was strange for me, as I've never really seen myself as that much of a people person.

So I arrived back in Cairo by train the next morning and spent my last few days in Egypt either sightseeing or helping Tahany in her lab in the evenings before returning to the Bedouin to sleep.   One afternoon I returned early because I'd left the guidebook in my room only to find Tahany's brother had bug-bombed all three floors, which meant I had a decision to make.   I could either go without my book for the rest of the day or run upstairs with my shirt covering my nose, find my room through a thick blanket of fog, grab the guidebook, and then head back down all in one breath, which turned out to be the toughest thing I did the rest of my time in Egypt, doubtless something I would have been unable to accomplish the night I bought the ticket back in Montana.   And I was barely out of breath afterwards.   One of the most consistent findings in medical research is how good regular exercise can be for the body.   Exercise is one thing that never seems to get "recalled" by some other study later on.   My last evening Mustafa gave me a lift to the airport and within a couple hours I was headed towards Germany the way the crow flies.

LET THEM EAT CAKE

As the plane gained in elevation, then veered off towards the northwest, finally leveling out above the clouds, I reclined further into my seat, trying to sleep but soon realized that wasn't going to be possible, so I pulled Gamal's manuscript from my pack.   Even unbound it was considerable and I wondered if I'd ever feel passionate enough to write that much about anything, including anthrax, or Montana, or Lewis & Clark, and so I began reading where I had left off down in Luxor, with the yeast having gained a culinary foothold in Europe.

Even though on separate continents, Cairo and Munich share some interesting similarities.   Both were at a crossroads, geographically – Cairo where East met West along the Silk Road, Bavaria in the heart of Europe, in close enough contact with the northern parts of the continent while at the same time still near the trading routes of the Mediterranean – providing Munich with a constant influx of new travelers with new ideas from different directions.

The Roman Empire's fall in the west marked the beginning of the Dark Ages in Europe, a period which was to last about a thousand years.[lxxiii] Gamal had a theory that the invasion of Rome by Germanic tribes from the north was due at least in part to the climate getting cooler, not to mention a desire for Roman wine.   Northern climates don't allow grapes enough time to make all the sugars necessary to impart a sweetness to wine so northern civilizations tended to become beer cultures, as is obvious today. Who likes sour wine?[89] I knew I didn't. Grains have a better tolerance for the cold and so it's no coincidence we have three microbreweries in my hometown but not a single winery, and Montana being the second largest producer of barley in the United States helps out too, I imagine. Northern grapes tend to be sourer; perhaps the climate got colder and the barbarians simply had a sweet tooth is what Gamal was implying. What is certain is that, rather than wither on the vine, viticulture became more widespread after the Roman Empire declined, when their technology for winemaking would have spread further afield.[lxxxiv]

Religious authority stepped in to fill the void left by the collapse of Roman order during the Dark Ages and monasteries needed to keep a constant supply of wine and bread around for the Eucharist.[90] In Germany, beer was one of the foods monks were allowed to take in for extra calories while still adhering to fasts. Like the roiling of carbon dioxide bubbles produced during a good fermentation, for many in the Dark Ages life remained in a constant state of flux too and ordinary folks just didn't stay put for very long. It's one of the reasons we have so few records of them today. [lxxxv 91]

I had selected a new guidebook – this one on Bavaria – while in a gift shop in Cairo and to be reminded of Bavaria's connection to monastic brewing I needed look no further than the full-page ads placed by brewers like Augistiner-Brau and Paulaner. Indeed, even the name Munich comes from the German word for monk. [lxxxvi]

But brewing was hardly new to Europe following Rome's fall, in fact the Celts had been making their own version of a beer called curmi on Britain long before. They also made mead by mixing honey and hard cider.[lxxxvii] Brewing in England was one of the enterprises the king didn't tax and so each medieval village had its own ale-taster,[92] an official charged with keeping up local standards. As had been the case with Egypt, brewing and baking was most often entrusted into the hands of housewives, in fact being a good alewife could be reason enough for a man to ask for a woman's hand in marriage.[lxxxviii] England doesn't have the tradition of monastic brewing like Bavaria because King Henry VIII put brewing in private hands.[93]

---

[89] In fact, the French still sometimes add sugar to their grape juice to allow the yeast to make extra alcohol.
[90] Some nunneries also brewed beer.
[91] During the Dark Ages, there were no fewer than 500 monasteries brewing beer in Germany, the most prodigious of which were the Benedictines. The daily ration of beer per monk was one gallon.
[92] William Shakespeare's father was an ale taster. / The Roman emperor Julian wrote a poem 1,600 years ago in which he described Celtic beer as smelling "like a billy goat".
[93] Ale was sold in bottles at the Globe Theater during Shakespere's plays / Drinking beer would become so popular that water removed from wells by brewers would cause London to sink three inches beginning in the 1860's.

And it turns out even the layout of the typical European city today was influenced by monasteries, making these religious orders the equivalent of "embryonic cities".[lxxxix]   Within the confines of a Middle Ages monastery one would have found much of what you'd expect to find in a modern town today, including a brewery, bakery, place of worship, and boarding houses for pilgrims, workshops with seller's stalls, even a small hospital perhaps.   And it was in this relatively stable environment that monks began experimenting with different varieties of grape, improving growing methods by taking seriously for the first time what seems obvious to any rancher worth his salt today...the quality of the soil and the amount of sun and water their crops received. Even the clarity of wines was improved upon in monasteries by adding egg whites to remove impurities.   Monks examined the connection between climate and vintage and experimented with grafting more productive shoots onto sturdier but less productive rootstocks in an effort to increase grape yields.   It was within the confines of the brewing monasteries of Germany that hops would be added to beer for the first time in the 11th century, which not only imparted a distinct bitterness to the beer, but also increased its shelf life by discouraging spoilage-causing bacteria.[xc]   This natural antimicrobial property in hops allowed beers with less alcohol to be stored for longer periods.

The monk Dom Perignon has been credited with inventing champagne but according to Gamal this misconception began with an advertising campaign in the 1800's.   In reality, the monk spent most of his time trying to eliminate bubbles in wine, which he himself found excessive.   Perignon did introduce innovations that would someday make champagnes more practical, though, like thicker glass bottles along with a rope to hold the cork in.   Interestingly, according to Gamal these French sparkling wines came about by accident after a cold spell one fall, which caused the yeast in the developing wine to become dormant and stop fermenting.   This was then followed by an early spring, which allowed the hibernating yeast to awaken and rapidly resume fermenting full force all the leftover sugars still in the grape juice, producing a sudden burst of carbon dioxide, shattering the bottles and creating the first Champagnes in the process.

Looking out my oval window I could just make out the jagged tops of the Alps towards the west, and I wondered if I'd been wise enough to place my passport near the top of my backpack, like an experienced traveler would have.   When ancient Rome fell it fragmented, eventually giving rise to many of the countries in Europe today.[94]   I skipped the airline's breakfast and went back to Gamal's manuscript.

---

[94] A more recent version of what occurred when the Roman Empire fell was the fragmentation of the Soviet Union in the 1990's, doubling the number of Eastern European countries in less than a decade.

Incredibly, it was the monastic thirst for improving winemaking and brewing during the Dark Ages that helped lay the groundwork for the *scientific revolution* that lay just ahead. It was this need to spread information about crops and winemaking and other manufacturing processes that led to some of Europe's first guilds, scientific societies, and journals. The typical tavern owner in England was also a vintner and likely benefited from this spread of information obtained by monks and the winemaking guilds.[xci] In the Middle Ages wine was added to drinking water during times of siege to help ward off disease and it's been recorded that King Edward II of England provided 4,000 barrels of wine to his army during one particular siege.[95]

I flipped forward through Gamal's manuscript until I came across another footnote, this one on bread, surprised to find that baking bread wasn't all that popular in Paris until the 1700's. Northern Europe has a much longer tradition of eating meat than baking bread. The practice of leavening dough did catch on when its time came and the role of the yeast changed as it spread throughout the Mediterranean region, radiating outward from Greece. In the Dark Ages, people didn't use plates but instead ate off of thick slices of bread called *trenchers*. These soaked up the grease and were then donated to the poor or thrown to hungry dogs after a meal. Baking was slowly becoming an industry in medieval Europe; however bakers still got most of their yeast in the form of skimmings (leftover yeast) from breweries. It wasn't until the explosive growth of cities during the Industrial Revolution that bakers would be forced to turn to a more reliable source of yeast.[96]

Bread's scarcity, its high price and poor quality, became a direct cause of the French Revolution and Gamal claimed that the hungry mob storming the Bastille prison in Paris was as intent on finding flour to make bread with as they were obtaining gunpowder. The French king made the unfortunate mistake of deregulating the sale of flour after several poor grain harvests so there was hoarding, which all combined like the perfect storm to create a massive bread shortage. Incredibly, a single loaf could cost as much as a month's wages, so it's not surprising that bakers in Paris accused of hoarding flour, or of making bread with spoiled flour, could be seen hanging in the streets following food riots.[xcii] Louis XVI's coronation took place at the height of a bread shortage, which is when Marie Antoinette was supposed to have said, "If they have no bread then let them eat cake," though there's no contemporary evidence for her saying it.[97]

---

[95] When Mark Twain toured a castle in Heidelberg, Germany he described a wine cask that held 221,726 liters of wine (the Heidelberg Tun) as being as big as a cottage and usable as a German dance floor.

[96] In 1871, a factory in Austria was the first to produce yeast just for bakers. Bakers today use strains of yeast more efficient in carbon dioxide production and with more *invertase activity* to cleave the sugars found naturally in the flour.

[97] What we call cake today is an enriched bread containing milk, eggs, and sugar.

My plane landed on time but I still missed seeing Frankfurt itself, blaming it on the efficiency of the German train system because it turns out there's a station in the basement of the airport, which is where I bought my ticket from a vending machine. Soon my backpack and I were being whisked through the rolling hills of southern Germany on towards the Alps of Bavaria with only a quick change of trains in Mainz first. I measured our progress by the gradually increasing slope of the landscape as it transitioned from flood plain into alpine forest.

Equally as important as the monasteries were to brewing in Bavaria was another act of divine intervention...the local geography. The weather patterns and soil profiles in Bavaria are nearly ideal for growing beer's ingredients, barley and wheat, not to mention hops. An area larger than either Ireland or Portugal and situated near the Alps, Bavaria has the advantage of being fed by rivers of melting snow with little limestone in the soil as this mineral has the nasty tendency of turning good beer cloudy.

Perhaps I shouldn't have been surprised to find that my Bavarian guidebook had devoted an entire chapter to brewing, explaining how it was the natural ice caves of the Alps (and later the Bavarian's attempts to imitate the year-round coolness of these caves by digging beer cellars near breweries) that would bring about the most significant change in brewing in thousands of years...the birth of the world's first lager beers.[98]

I held the colorful pictures of Munich's beer halls up to my nose, taking in a whiff of fresh ink from the pages, the sharp stinging helping ward off my fatigue for a moment before I continued on. Unlike much of Europe, Bavaria remained stable, politically, for a very long time. The kingdom was ruled over by the same family for more than 700 years and has the world's oldest food purity law to show for it.[99]

As my train rounded another bend nearing the outskirts of Munich, I couldn't help wondering if Wolfgang (or Wolf, as Gamal referred to him) had any intentions of honoring our agreement or whether he would leave me sitting in the Hofbrauhaus. The retired microbiologist apparently had no mobile phone or any other way to get in contact with him, which made me wonder how Gamal had been able to keep track of him. In fact Wolf, as far as I could tell, didn't even use a computer, which is virtually unheard of even for a retired professor these days.

---

[98] The coolness of the ice caves increased not only the shelf life of beers but also their clarity, producing the amber color which makes most ales look like used motor oil from my pickup truck by comparison.

[99] Reinhettsgebot Law – first proclaimed in 1516 by Bavarian Duke Wilhelm IV – said that only hops, barley and water could be used to make beer.

"Somewhat of a character," was how the diplomatically inclined Gamal had described him.[100]   There was even a list of things I wasn't supposed to bring up to him.   At the very top of it was his forced retirement.   In Germany, professors faced mandatory retirement and apparently Wolf felt he'd gotten the shorter end of the stick.   The thinking is that scientists do their most productive work by age 45; then one should make room for the younger scientists coming up.   It certainly seems true enough, sometimes.   Einstein did his most significant work as a young man even before he found his way back into academia, as a newly minted Ph.D. just out of college while working at a government patent office. [xciii]

I propped my glasses on my forehead so I could continue reading but still glancing out the window once in a while to gauge my progress.   After skimming through the section on breads, my stomach began a low but audible rumble.   It seems that at first monasteries were making beer for their own needs, but King Ludwig eventually changed the law so lager beer could be made available to the public.

Ludwig realized he could tax beer and so it wasn't long before Munich's outdoor beer gardens began to dominate the city and Munich's poorer residents found they could bring their own food and eat under the shade of chestnut trees (the trees did double duty by keeping the underground cellars dry, absorbing the rainwater with their roots).   This picnicking resulted in a crisis when brewers realized how many profits they were losing from the sale of food, so the King stepped in and hammered out a new agreement and as a result the poor would continue to bring their own food to the beer gardens, but could eat only at uncovered tables, which can still be seen today as the self-serve tables of Munich's beer halls.

I caught the unmistakable whiff of charred meat from a vendor's wagon through an open window while flipping pages in an effort to find out where Hofbrauhaus was exactly.   That's where I was supposed to meet Wolfgang and the only information I could find was a warning about a "touristy feel", but the upside was that the beer hall was guaranteed to be like Oktoberfest any day you drop by.   Hofbrauhaus became a brewery in 1589, and then like so many beer halls in Germany, it became an institution.   Entire families made a day of visiting beer halls here and have for generations, still bringing picnic lunches to the gardens.   This is, after all, a city with a college that doubles as a brewery for teaching Europe's next generation of brewmasters the ancient art.[xciv]

---

[100] Gamal had mentioned that Wolf made a name for himself as a successful "yeast breeder" earlier in his career and that he came up with the first strain of yeast capable of surviving in up to 25% alcohol, but apparently failed to cash in on the recent 'extreme beer' craze, microbreweries, and home brewing.

THE YEAST MASTER

I left the train station on foot barely aware of the heavy load over my shoulder and so it didn't take me long to find the beer hall. The train had dropped me off in the heart of the city – as is true with most trains in Europe – and I simply walked the rest of the way to Hofbrauhaus. The first thing I noticed near the hall was a costumed blonde waitress about college-age walking out the door carrying a tray, charged with the task of delivering two pitchers of beer through a crowd of tourists. To my surprise, the pitchers would turn out to be drinking glasses and after entering and locating a seat near the stage, giving my eyes a chance to adjust, I was able to see that the place was filled mostly with tourists hoisting beers while someone with a smartphone always seemed to be angling for a photo. On the stage closest me were two Alpine singers who could have been cast members from *The Sound of Music* – a man dressed in black lederhosen chopping wood with an actual axe while next to him stood a woman, unconcerned and adorned in traditional dirndl dress and broad red apron.

German beer has a reputation for being strong, apparently because it's purer, often having been brewed on the premises and so without any preservatives, which means it has a better taste. And served in larger portions means many first-timers like myself simply make the mistake of drinking too much. The guidebook had neglected to mention this, plus the fact that I was drinking on an empty stomach meant a hangover was in my future. The table next to me had a plate of fresh bretz'n – soft pretzels – heaped on top of one another, but when the waitress came by it slipped my mind to ask for any even though I was hungrier than a woodpecker with a sore beak. It seems I was still on autopilot, nervously searching for the elusive Wolf. In fact, I had begun to suspect he didn't really exist, or more likely that I'd missed him and he'd stormed off in a huff and was already phoning Gamal at that moment. Giving up, I began scanning the hall for the nearest restroom, the image having popped into my mind of the great Danish astronomer Tycho Brahe, the one who had helped Johannes Kepler with his calculations and how in 1601 had met his unfortunate end at the hands of a burst bladder while drinking too much beer at a festival in Prague.

In spite of the hall's enormity, I was the only one who seemed to be alone at a table and so with an oversized beer as my companion I pulled the guidebook on Bavaria back out, sipped the foam off the top of my glass, and picked up where I had left off, reading once more about Munich's beer gardens.

The book claimed that some beer gardens could hold 7,000 drinkers, about twice the summer population of my hometown (not including the university students who always arrive in the fall). [xcv] Some of the more established ones have chestnut trees so full that their tops can keep customers dry even during a downpour. But it's not just about drinking since beer gardens are a part of everyday life in Bavaria. Some even have petting zoos and playgrounds for children while others may boast a small lake with a cycling path.

I had made it halfway through my liter ration of beer when a cold sensation came over me. Looking around, it felt as if everyone in the hall had dropped what they were doing and were aware of only me.

But before I could turn completely around I heard in a thick German accent: "So you are the one who thinks he can learn the ways of the yeast in only two weeks."

I swiveled around in my seat and there towering over me like a shade tree was a very large man. My first thought was that he must be listed in the Guinness Book of Records, or at least was in the running. He had pale eyes surrounded by sunburned skin and the sturdy frame of a much bigger man than Gamal had led me to expect. His hair was straight, long, and thinning at the temples but gathered into a ponytail neatly behind his head. The beer hall was dark and he was still too far away from me without my glasses on to make out the tiny blotches of broken capillaries that would come to remind me of spider's webs on his cheeks; the remnants of too much sun on trips to the south of Spain as a youth perhaps. He was wearing faded blue jeans sprinkled with small holes, and a t-shirt beneath a blue-jean jacket, the jacket matching his pants in its degree of fadedness. All in all, not exactly what one expects to encounter in a retired microbiology professor.

"Wolf?" I asked, gingerly extending out my arm, forgetting to ask if I could call him by his nickname, "Gamal has told me a lot about you."

He reached out and suspiciously pumped my hand while I wondered why Gamal hadn't mentioned his size. Most people would have mentioned someone this enormous when describing them.

"Gamal says you are eager to learn all the ways of the yeast. But the real question is...do you believe it dock-torr?" With one hand he guided the wooden chair opposite me out from beneath the table and slowly sat down. There would be no need for further formalities. He motioned for the waitress to return, then leaned back and stretched his legs out in front of him, crossing his ankles. I could see he had tennis shoes on but not bothered with any socks. Leaning further back still, he placed his arms behind his head, interlocking his fingers to use as a headrest. He did this all the while keeping his eyes trained on me. My initial impression was of a man who couldn't be bothered with what my impression of him might be; it would only be his that mattered. I wondered how Gamal had managed to find him, let alone gotten him to sit for all those interviews. He said Wolf had been a brilliant scientist at one time, before...whatever it was I wasn't supposed to ask him about.

"Yes," I wanted to reply, "I'm just starting out with the yeast. Not really a eukaryotic scientist, and definitely not a fungi expert by any stretch. Gram-positive bacteria are more my speed," but what actually came out was "Um...well...I did work with *Bacillus anthracis* in graduate school," as if I were somehow ashamed of it.

"That causes anthrax," Wolf blurted out and an elderly couple, a man and a woman, turned around in their seats to look at us as soon as they heard the word anthrax. It's a strange sensation even for a scientist to hear someone talking about a disease in a beer hall with an Oompah band in the background.

Trying to regroup, I added, "You come highly recommended. Gamal says that if anyone can bring me up to speed, you certainly can."

"Ah, little Gamal. Well I hope you are up to the task. He seems to think so, but I'm afraid I already have my doubts." The seriousness in his tone made his words feel more like a diagnosis and the microbiologist then began strumming his fingers on the table. He was still staring at me.

"The yeast is not so easy. You are finding this out. And time is short, yah? You are not so young anymore, are you dock-torr?" He emphasized the "dock" half of the word, giving it what I felt was an unnecessary air of sarcasm. Playing possum, I decided to take the role of listener, assuming that is probably what Gamal had done much of the time he sat across from him. Just lay low and keep an open mind is what the little voice inside me kept saying. Besides, they say that in spite of appearances, Germans actually enjoy their pessimism and that no foreigner has a right to deprive them of it.

"I worked with anthrax in grad school, that is until," I listened to the sound of my own voice fade, easily replaced by the band, then leaned my face down towards the top of my mug, feeling what was left of my confidence drain into it. 'Was he really implying I couldn't handle the brewer's yeast', I wondered, finishing the rest of my beer.

"Yes, I know. You used to verk with bacteria," his face crinkled as if saying the word left a bad taste in his mouth. "Gamal has told me."

I immediately thought back to my conversations with Gamal in Cairo, trying to recall if I'd ever mentioned anything about losing the anthrax grant. No, I was pretty sure I hadn't. But I did mention it to Tahany in one of my e-mails right after it happened, I was pretty sure. Yep, sure enough. Wolf knew too. I tried to take another sip from my empty glass as the waitress returned with two new ones. At least someone was showing good timing.

Cradling his beer, Wolf, his elbows planted firmly in the middle of the table, advanced his upper body towards me, his voice directed into my ear closest to him, speaking as if he were letting me in on a secret.

"The brewer's yeast may not mean much to you. She is not so pretty to look at. She does not move around by herself. Why? She has no means. Yet she harbors more secrets, possesses more talents, than you nor I nor anyone else will ever know, dock-torr. She has everything to teach us if we will only listen...take the time to examine...ask her all the right questions. She is not an *E. coli* and you don't simply dab a toothpick into some sterile broth and feed her some sugars. You must remember if you want anything worth publishing. You don't streak her onto a Petri dish, lock her away in some incubator and then forget all about her while expecting she will reveal everything in a single afternoon in between teaching science to children. This isn't a game dock-torr. This is a dialogue...a dialogue between you and one of evolution's greatest creations."

He spoke as if the yeast were someone he knew personally and felt the need to give it some sort of a character reference. Pretty strange, I thought, but then again, if I hadn't learned anything from all those years of college, it was that microbiologists can be an eccentric lot. So far, this one had them all beat, though. Wolf went back to leaning in his chair, apparently content to study me from a distance again, hands interlocked behind head and using questions to probe me the way scientists do an experiment, gauging my reaction after each perturbation of me.

I tried to sip some foam off the top of my glass, the foam itself having collapsed and was now trying to retreat down the side of my mug, my chair becoming larger, the same chair that just a moment ago seemed to fit me fine. I decided to give the conversation another chance to develop in a meaningful direction. My tenure might just depend on it.

MULE OF THE ICE CAVES

Only when the music lowered enough and the yeast master felt sure he had my full attention did he continue. "Well, I will begin by explaining that it is YOU dock-torr who is the student and the yeast is the teacher. Is that clear?" It wasn't really a question so I didn't answer.

"This..." he proclaimed while holding up his mug, "...is a lager. The pretty yellow color you see here...the fresh taste...it would all not have been possible if not for diss humble fungus. A microbe that gets no respect. Hundreds of years ago, not far from where you and I are now, diss tiny microbe went its own way...changed and the world would never be the same. It changed much more rapidly than any bacteria ever will. Have you ever seen a mule dock-torr?"

Of course I'd seen a mule. I grew up on a ranch. Still, I just shook my head no. It seemed the safest thing.

"Well the mule is a hybrid...what happens when a horse and a donkey mate. The same thing happened with the brewer's yeast inside an ice cave high up on the mountain," he gestured towards the Alps, "It came in contact with this other yeast, a different species, and it didn't know what to do so it went ahead and did its own little experiment. It mated with it." The audience let out a loud gasp, then a spontaneous roar of laughter as I craned my neck towards the stage just in time to catch sight of a fully-grown man in lederhosen trying to plant a kiss firmly on the cheek of a rather plump woman with red cheeks. He failed and looked on forlornly as she ran off the stage giggling. The microbiologist waited for my attention before resuming.

"The yeast accomplished its task, and yet it failed at the same time. Have you ever had an experiment give you more than one answer depending on how you looked at the results dock-torr?"

Once again, I didn't answer other than to just nod my head yes this time.

"This produced a hybrid...something we would call a lager yeast.[xcvi] This new hybrid yeast, this mule of the ice caves, she came about by happenstance, and yet she could ferment beer in the cold at the bottom of the vats in the ice caves better than any yeast before. And it's all because the yeast can have sex. Can your bacteria claim that? No, I don't think so. Certainly it was not the haphazard random kind bacteria might boast of once in a while when they shed a small piece of DNA and give it to another, but the true kind when two cells merge to become one, not so different than how human cells come together at conception you see?"

"I worked with anthrax, not *E. coli*," I added, looking for signs he even heard me above the band.

"Well this new yeast, this mule of the ice caves, it produced this vunderbar new beer with a clearer, crisper flavor, not the complex profiles of those dusty ales, the ones the British seem so fond of. Mine got I will never understand diss," his accent seemed to wax and wane and I soon found I could use it as a gage to his emotions, "You know when we first began adding hops to beer in the 11th century, the British on their island way up there, they actually thought we were trying to poison them? I sometimes think it is the reason those limeys still drink ales...and they drink them warm, too," he made a sour face, "They still think we want to poison them, I suppose." [101]

I found it amusing how he seemed so obsessed with my having worked with bacteria before. To him a yeast cell on its worst day was much more clever and interesting than any bacterial cell having the audacity to lack a nucleus on its best day. [xcvii]

"You know, because of our lager beers, people have always come from all over Europe to study brewing in Munich. The man who started Carlsberg in

---

[101] The English began using hops in their beer around 1700.

Denmark...Jacobsen was his name...well he came here too, and he studied beer like all the others. And it was in Munich that refrigeration was used for making beer first. [102] You probably didn't know that either. No need for ice caves anymore," Wolf's voice became calmer the more he talked of Munich and its unique contributions to brewing. He then pointed to a clear drop slowly dribbling its way down the side of his mug.

"You see this dock-torr? Each drop it takes a million yeast cells to make. Talk about your slave labor. They are so amazing, though," he paused while admiring the droplet rolling down his glass and I half expected to see a tear form in his eye. "They do fermentation not with DNA but with proteins. If you dry out a yeast cell almost two thirds will be protein," then he erupted "So, why do you think there are so many breweries in Wisconsin? Certainly you must have some answer. It is where you are from after all."

"German immigrants, I suppose? Actually, I'm from Montana, Wolf. It's a different state entirely."

"Yes, of course...because Germans went there, but what else?"

I thought for a moment, then gave up and only managed a puzzled look.

"The lakes...with their ice of course. To make good lager beer like in Germany, they needed winter ice...to recreate the coldness of the brewing caves of Bavaria...but in Wisconsin...or is it St. Louis?"

As if a consolation prize, the waitress returned and placed a plate of soft, salted pretzels in the center of the table, still steaming, and it occurred to me I was halfway through my second liter of beer without having eaten anything since Egypt. Wolf picked up one of the pretzels and began studying it in his hands like a bug. I was relieved whenever his attention was drawn away from me, and yet I braced myself because I sensed he was about to become sentimental again.

"Soft bretz'n. Bread made with handles," he cooed fondly over the golden-brown, leavened pretzel while turning it over in his hands, caressing it almost, his hands so large that the pretzel almost disappeared, the bread perhaps reminding him of his own Schwabish childhood. "Pretzels like this could be hung on hooks, or exchanged between knights on horseback...with just the tips of their spears. They were invented when a monk used them as bribes for the kinder, so they would memorize their prayers.[103] See how this one it looks like a child praying...with the arms crossed?" He held it up to the light and I scooted my chair in closer to the table, having trouble hearing him now that the Oompah band had picked up additional members. "You see these holes? They represent the Trinity."

Sometimes Wolf spoke so calmly and quietly he seemed almost normal and yet his demeanor could change out of the clear blue, so I didn't want to

---

[102] Spaten brewery, 1873. It allowed for the first year-round brewing of lager beer.
[103] AD 600's.

miss anything if he should say something important about the yeast. I sensed I could learn a thing or two, if I could just get past his exterior. There was no sense in asking questions, though. I knew that. It's always best to ride a horse in the direction he's already going. A little braver I decided to test the waters of this latest mood.

"You know, bacteria are pretty complicated too despite their small size. It was, after all, someone from your own country, Robert Koch, who worked out anthrax's lifecycle...of it alternating between germ cell and spore."

"Yes," Wolf replied wistfully, still fondling the pretzel, "he was a very detailed, careful thinker our doctor Koch," then he looked straight at me and erupted, "Not like these dummkopfs today, these gene jockeys with their enriched media. Do you know they actually bragged they will make paleontology obsolete someday? Mine Got! The entire field of paleontology! Obsolete! Just like that," he snapped his fingers, "with only DNA! More publications, more and more, all the time, that's all they think about, Christ why is this, and they don't care how they get them, either," he pounded his fist on the table without harming the pretzel any, "And bringing in grant money, traveling to symposia...that's another thing, these are the ones who are the bane of good science. These dummkopfs with their machtpolitik."

I clearly had touched a nerve and backed off, hoping he had gotten it all out of his system, considering myself lucky not to have been the object of his wrath. Mark Twain once claimed that whoever invented the German language must have done it after staying up all night with a toothache.[104] After regaining his composure, Wolf went back to describing how proteins from cells like the brewer's yeast, and human cells as well, have been relegated to what he called the "Stief kinder", or stepchildren, of the life sciences, while accusing DNA of grabbing all the headlines, of being a "dead molecule in disguise of a living one" as he put it. [xcviii] I couldn't argue with him too much because on one level I actually agreed. The DNA chromosome is mostly a storage place – a repository if you like – for information, when compared to proteins. In spite of its length, DNA is a relatively modest molecule, both physically and chemically compared to proteins, which is why DNA can be recovered from mummies and Neanderthal bones even after thousands of years and still retain useful information. It's simpler. Proteins on the other hand are the most "alive" molecules in our bodies. They are the closest things we have to robots working away deep inside us. It takes proteins, for example, to make a chromosome divulge all its information to the cell, or to make another copy of a chromosome during mitosis, or move about if you're a muscle cell, or to turn sugar into alcohol if you're a yeast. But at

---

[104] Twain also claimed that German words were so long, they had a perspective.

least I'd made my point and that was enough…for now. I silently chalked one up for the bacteria. [xcix]

Cooling somewhat he then added, "As a prokaryotic person you are no doubt familiar with the Dutchman…van Leeuwenhoek then, eh?"

"The first person to lay eyes on bacteria," I answered confidently.

"And the yeast. Do not forget about the yeast. Leeuwenhoek's family on his mother's side they were all brewers. Many people do not know this. He was interested in brewing because brewing vas in his blood."

"I'm not too surprised. Wasn't even Fredrick the Great a brewer?"

"Yes, but right now we are discussing the Dutchman," he looked away as if disappointed and I silently resolved not to change the subject anymore.

After a slight pause, Wolf leaned in closer and began again, "He was the wine-gauger of Delft even though Delft had a reputation as a wunderbar place for a beer. It was Leeuwenhoek's job to go around and visit all the merchants and measure the wine…for taxation purposes. So anyways, this Dutchman, he makes these perfect tiny lenses in his spare time until one night he gets the idea to have a look at his beer, and not the finished beer, either. Oh no. He is a true scientist. He wants to know what is happening while his beer is still being made…during fermentation.   So he puts a drop of sweet barley wort on his microscope and he fumbles around with the light from his candle. He has hundreds of handmade microscopes by the way. He made a different one for each microbe he wanted to look at.   He was a very self-reliant man, diss Dutchman, not like these spoiled kids today. Even a music student can pop open a catalogue and order a microscope these days. It does not take that much…but to be a good scientist…ah that dock-torr…is another matter. So Leeuwenhoek he sees the yeast, but he mistakes them for pieces of starch from the barley.   Mine got, as brilliant as this man was…he simply missed them you see?   Yeast do not swim around.   They do not have to. So they just sat there in front of him and he was fooled into thinking they were nothing.   And as a result, the yeast would keep its secrets for 150 years more. By the way, why do Canadians wear those flags? The ones with the red leaf on them.   They sew them onto everything, their backpacks, their shirts, even their luggage. Mine got, I even saw this Canadian once who had one on the seat of his pants."

"I don't know. Why are you asking me?" I was dumbfounded.

"I just thought maybe since you are from America, and Canada is right next to America, you might know."

"Pride?" I suggested, "It's just a guess."

His question gave me a sense of vindication for who was changing the subject now?   Wolf picked up his beer, wiping his mouth from one corner to the other with a sweep of his sleeve and then began looking again into the golden liquid inhabiting his glass, clearly taking delight in it before downing another continuous gulp.

"He was a very curious man, this Dutchman, and he wanted to know what caused all the bubbles in his beer, that is all. Curiosity. Why is just finding out the answers not enough anymore? He didn't do it for the fame, although he did become famous. And he didn't do it for the money, either. No. He only wanted to know if any of his little animalcules could live without air. Yes, he found out, they could. But the Dutchman, he missed the fermentation part because...well because he didn't see the yeast as alive. Mine got he simply didn't see it even though they were right in front of him,"

Wolf slammed his mug down on the wooden table and it made a hollow ringing sound. The same couple as before turned around and stared at us again while I scrunched down in my seat a little further. His accent increased in thickness but then thinned out as he was capable of calming just as quickly.

"But he did scrape some bacteria from his teeth and he found he could kill them with heat and vinegar. So Leeuwenhoek he becomes not only the first person to see bacteria, but also the first to kill them. You should know."

I didn't reply even though I knew what he was saying was true. Wolf wasn't the only microbiologist with an appreciation for history. They didn't need government money to run their labs in those days, which was why I currently had an affinity towards scientists like Leeuwenhoek. Most scientists back then were wealthy, or had sponsors, and governments didn't seem to feel the need to support research much in those days.

"Yes," Wolf said wistfully, leaning back in his chair, "He invented a mouthwash too...just a mixture of vinegar and wine. That's all it was. And diss Dutchman he had over 400 microscopes he made with his own hands. Did you know?" Wolf was working on his third beer while I was still nursing the foam from mine.

"Yes. You said something about that, Wolf."

We sat without talking for a while longer, perhaps ten minutes more, my attention focused mostly on the band but not really listening to them. It seemed the right thing to do as they were putting an awfully lot of effort into whatever it was they were doing. Feeling higher than a Colorado pine tree, I had little idea what Wolf was thinking and didn't try to guess anymore. A few minutes later and the yeast master picked up where he left off and as was becoming my habit I took up my position, leaning forward towards the middle of the table, my cheeks cradled in the palms of my hands in an effort to hear him as the band had begun playing another merry polka.

"The yeast kept its secrets so long because it likes to wait, you see? Now that is patience. A single yeast can ferment its own weight in sugar every hour if you let it. You must remember diss. Imagine yourself eating 70 kilograms of sugar in one hour. Could you do that? I don't think so. Nor could I, nor could anyone in your department back in Wisconsin."

"No," I agreed, wondering if the pretzels were mopping up the beer in my stomach, "I'm from Montana."

"Well the yeast can do the equivalent. Eat it's own weight every hour. You cannot imagine and barely can I. Well when you can, you will begin to appreciate all the wonders this little creature holds in store. So it was left up to a Frenchman to take up the challenge...of proving it was indeed the yeast behind perhaps the greatest miracle of all time...of fermentation. It had baffled so many all the way back to ancient Egypt. Diss is what I could never get Gamal to understand, the importance of asking questions just for the sake of asking. Well, the next person was Louis Pasteur and he was a chemist by training, but he wasn't a coward and he wasn't afraid to change fields. Maybe a bit like you that way."

My first grudging bit of respect from him and I wondered if he was warming again. The fact is, I'd known about Pasteur, and about how, as a young professor, he had found himself in a remote French village teaching chemistry in the mid 1800's, which is also where he began his historic work on fermentation. Leading chemists of his day thought fermentation was strictly a chemical process so they assumed there was no need for a living organism. I identified with Pasteur too...as he was under some of the same pressures I was now, to prove that scientific work can be meaningful in a practical sense. That's been the case for a while and I expect it will continue. Wolf was right in a way. Just finding out the answers isn't enough anymore. We need practical reasons for doing our work now. There's even a place on your grant application where it's a good idea to try and answer this if you want your grant approved.

"Anyway, Pasteur he turns into a biologist and decides to use some of the same skills he learned as a chemist in Paris to determine once and for all why it was that grape juice had this tendency to turn into wine. You must understand, the French are always more concerned with their wine...more so than their beer. They are French, after all. They cannot help it. It is in their nature. Well as a lecturer in Lille, he became quite well known. Lille was an industrial town so one day a boy whose father owns one of the sugar beet factories for making distilled alcohol, he asks Pasteur to find out why his beet juice keeps going sour. It was costing the family a fortune each time a new batch of beet juice spoiled and had to be thrown away.[105] Pasteur was a vinophile so he knew about the same problem in wine, in fact he sometimes returned to his family's farm to relax.[106] They had a vineyard there and

---

[105] The French were using sugar beets rather than sugarcane because of an embargo by the British in the West Indies.
[106] So did Pliny the Elder, who wrote about diseases of wine almost 2,000 years earlier.

Pasteur he walked around it in the evenings, thinking about fermentation, I suppose. He even brought his microscope home with him, perhaps much to the displeasure of Mrs. Pasteur, I don't know. So he decides to help the boy's father which means taking his microscope into the factory to examine the vats of beet juice up close. It does not take him long to realize that in the vats that keep going sour, he finds this tiny rod-shaped bacteria...millions of them. But in the good vats that produced the alcohol, he finds only the yeast. And not only this, but in the bad juice the Frenchman he sees that the yeast cells are becoming deformed, changing into strange oblong shapes, in the sour juice, the ones with the bacteria...as if the yeast are becoming sick or something. He was quite observant for a Frenchman. And of course being French he had an imagination to go along with it. What Pasteur saw was a kind of *war* taking place within the beet juice...it was a battlefield between good versus evil. Whoever won got to make their product."

"To the victor belong the spoils," I added, confident the pretzels were doing their job and so I hoisted my mug and took a dent out of my beer.

"If the bacteria won, then the product would be vinegar. If the yeast won it would be alcohol. It's how Pasteur became convinced of germ theory," I added with a certain triumph, "The notion that certain diseases, infections in the body for instance, were caused by bacteria. No one had any good evidence for this before."

"Yes. And the brewer's yeast led him to it, you see...that bacteria can cause diseases the same way they ruin good wine. Pasteur astonished the people of Lille because he got so he could even predict ahead of time, like a fortuneteller, which vats would go bad, and which would not, just by looking at a single drop of juice under his microscope. Mine got he was a true scientist."

"He eventually found out what Leeuwenhoek knew before, to use heat to get rid of bacteria," I added.

"And this is why the first use of pasteurization was to preserve wine, not milk.[107] It became so widespread that the emperor Napoleon, the third one, what's his name, he wanted Pasteur to use this new process to solve the problem of sailors deserting when the wine went bad on ships, which, of course Pasteur did." [c]

Wolf continued, his voice now steadier and more even than before. It probably helped some that the band had gone on break.

"Pasteur's insights were not lost on the English surgeon Joseph Lister, who takes Pasteur's knowledge and he applies it to fixing compound fractures, the kind that always got infected and ended in amputation. So Lister he keeps the germs out of the wound using carbolic acid, one of the first

---

[107] 1886.

modern antiseptics applied to the skin. And it works, so he confirms germ theory and gives Pasteur the credit. How is that for finding a practical application for your work, eh dock-torr?"[108]

"Pasteur was a great man," I lamented.

"But there was at least one thing he failed to do."

"What is that?" I wasn't too surprised there had to be a catch as far as Wolf was concerned. Pasteur had the misfortune of, after all, being French.

"He never could prove how the yeast accomplished fermentation. He tried and he tried...for more than half a century he tried to break open the yeast cells...to isolate, chemically, whatever it was inside the yeast that accomplished this miracle. But he was never able to do it. No, it was left up to another chemist...a Bavarian this time...and his name was Eduard Buchner.[109] After much effort, Buchner saw a way to extract the yeast juice, its cytoplasm, from inside the yeast, and along with this spilled out the enzymes that could accomplish fermentation. Enzymes are of course proteins. It was like squeezing oranges to get the juice out. He used sand to grind open the yeast cells, scraping them this way and that, and then he pushes the broken cells through a filter using high pressure and so the clear juice that flows out it goes into the beaker, leaving behind the broken parts while capturing only the cell's interiors. When he adds some pure sugar granules from a jar on his shelf to this pressed yeast juice, he sees bubbles form, the same bubbles people had seen for so many thousands of years while making beer and wine. Only he was at this moment the only person who could make alcohol but without any cells. The very first. He did fermentation with just the chemicals...the proteins from inside the yeast, some water, and a bit of sugar."

"Buchner became the first *biochemist* by doing that one experiment," [110] I added.

"Yes. You do know your history, don't you dock-torr? But it only seems simple now. Buchner did this in eighteen-ninety-seven because he tried something new. He did what we would see as the very essence of biochemistry today," Wolf was patting his open hand on the table rather than pounding it as a fist like before, to more gently drive home his point. Then he began pointing his index finger straight at me and exclaimed, "He recreated a process that took place only inside cells before, but he does diss in a test tube. No bacteria. No yeast. No living cells of any kind. Just the proteins along with some sugar water. He was the first to do cell-free fermentation.

---

[108] Lister's father was a wine merchant & amateur scientist. *Listerine* mouthwash is named in Lister's honor and was the first over the counter mouthwash available in America, in 1914. The ancient Egyptians made due with salt as a mouthwash.
[109] Before beginning study at the University of Munich to become a chemist, he worked in a canning factory in Munich.
[110] He published his results in an 1897 paper entitled "Alcoholic Fermentation Without Yeast".

And it changed forever the way people would look at life, not just the yeast, but all of life by this one experiment." [111]

I had already known about Buchner and his brother, who was also a chemist, having taken a course in the history of biology as an undergraduate.

"It's also true," I added, "that this experiment helped end the debate that had gone on for years, the notion that life was too special to understand using chemistry or physics, that living things had a divine spark that could never be reproduced in any test tube."

"Yes," Wolf was becoming more excited so I braced myself. But it turned out to be for nothing, as remained calm, "You are the cleaver one, aren't you dock-torr? You know your history...for an American that is. This idea of vitalism, an idea that went clear back to the Greeks and Aristotle, that all living things possessed special forces that made them impervious to the same things ordinary matter is subject to. Made them impossible to do experiments on."

We both sat for a while, I for one contemplating all the implications of Buchner's work. How if it wasn't for the yeast and Buchner, how biochemistry wouldn't be where it is, how we might not have molecular biology and my new yeast project, which grew out of Buchner's choice of the brewer's yeast as a research subject.

Wolf was smiling now, no doubt savoring the contribution Bavaria and his beloved yeast made to the history of something he clearly cared about. I also knew from researching the grant that if my lab was fortunate enough to discover any new yeast proteins, we'd be depositing their sequences of amino acids right here, electronically, in a database called the Munich Information Center for Protein Sequencing.

"You know, some time after Buchner accepted the Nobel Prize in 1907, you know what he said don't you?"

"No," I replied, "I have no idea."

"He told everyone that the reason he had succeeded where the Frenchman had failed for so long was that he, diss humble Bavarian, had made use of the leftover yeast from the breweries of Munich...the same hardy lager yeast that preferred the ice caves up in the Alps. Do you remember? Back in France, Pasteur had been using yeast that made only ale...at room temperature. Kind of wimpy compared to what makes diss," he hoisted h s mug high enough to signal to the waitress for another beer. Class apparently wasn't over for the day.

"Sometimes the devil is in the details when it comes to experiments," I added.

---

[111] Buchner wasn't the first to work with proteins. At least 1000 years earlier shepherds realized they could cut open the stomachs of goats and use seawater to extract a solution (the protein is called *chymosin*) to make cheese from goat's milk.

"Pasteur died before they gave out Nobel Prizes, or he would have gotten one…for his work with the yeast," Wolf acquiesced.

"Or anthrax," I added, knowing that what I was saying was true, yet it was also true that I was thinking more about a plate of steaming hot bratwurst topped with sauerkraut that had just traveled by at eye level – more so than I was arguing with an eccentric German who could be a few pickles shy of a barrel, one who had obviously tossed back a few before he even sat down with me.[ci]

THE GLOCKENSPIEL

After the waitress returned, he downed his beer with another gulp, looked at his watch, and then demanded we take a walk to see the town's Glockenspiel. The room had been getting more crowded and the thought crossed my mind that it might be easier to catch a bite to eat from a street vendor than wait for the waitress. Before I knew it, we were out the door and a cool breeze sweeping down from the Alps was helping wake me out of my lightheadedness. Perhaps the pretzels were doing their job as well. Things were looking up as we headed towards the large public square called the Marienplatz. The plaza has been the center of trade in Munich since medieval times.

Along the way Wolf began helping me with some of the more practical aspects of the project while probing me with questions regarding my plans when I got back. He offered up a cornucopia of advice both practical and theoretical, sprinkled with the kind of details you won't get from any journal, including which supply houses to order our chemicals and yeast from, which houses to stay away from, as well as whom to trust and whom not to when it came time to share any preliminary results we might get. He apparently knew all the players in the field by their area of expertise including which university or brewery they happened to be at, even which proteins they were interested in, however he didn't always seem to remember their names, which at the time seemed odd, but maybe not so much given his penchant for beer.

With darkness descending and the lights being switched on inside the houses, the neighborhood felt like one of those Grimm's Fairy Tale villages and at one point we passed a gingerbread-looking building on the opposite side of the street that turned out to have special significance.

Wolf noticed my attention drawn to it and explained, "By 1750 there were 4,000 breweries here in Bavaria. Beer was safer to drink than river water. It still is.[112] Even in England in the 17th century, the kinder in boarding schools

---

[112] If not for its unclean drinking water 2 centuries ago, New York City wouldn't have grown into such an important brewing center in North America.

would take bottles of beer to school with them. That David Copperfield character from the novel, even he drank beer when he was only eight. You see that building across the street? It is the Bavarian National Theater. It caught fire in the winter of 1823...yet it was saved by beer not once, but twice"

"Twice? How's that possible?"

"The first time was to put the fire out. It was so cold in Munich that the water had frozen over, so they had to use beer from a brewery to douse the flames."

"And the second time?"

"To rebuild it, they raised the beer tax." There was silence for much of the time as we walked the rest of the way to the square.

"Yes," Wolf began again finally, his mind apparently more receptive in the open air. "We had peace for over 700 years here...and the first streetlights in Germany that ran on electricity, too. How strange to think that Munich would someday become the birthplace of Germany's greatest sorrow."

"Greatest sorrow?"

"I am speaking of course of Nazism. You see, after the First World War Munich along with the rest of Bavaria had fallen into political instability. It was terrible and it affected everything. You probably don't appreciate it Inflation was so terrible the bankers weighed out the money instead of counting it. German farmers in beer halls ordered two beers in case the price went up before they got done finishing the first one. There vas starvation in the streets and the people sometimes they ate rats just to stay alive. Most of the intellectuals had fled to Berlin, leaving Munich to the reactionaries. Ever Vladimir Lenin lived here for a while. For a time, Munich was a Soviet Republic. Well, as you can imagine, diss was a good nest for radicals, and one in particular who set his sights on overthrowing the Weimar Republic. In 1923, hoping to spark a revolution that would spread all the way to Berlin in the north, with his small army Adolph Hitler he walks through the doors of a beer hall here in Munich."

"I've never understood that, really. Why a beer hall? And why here?"

"I was just coming to that. You see, like most politicians, Hitler had developed his speaking skills inside Munich's beer halls. Beer halls have always been important, not just as places to drink, but to socialize," he emphasized the word socialize by saying each of its syllables more slowly. "To live life, to discuss issues of the day. These enormous halls were natural gathering places, as you just saw. They draw Bavarians like moths to a flame, and if you are a politician, where better to find a receptive audience? In America you prefer your soapboxes in public parks...well in Bavaria we have a fondness for our beer halls. It's that simple. The famous Bierhallenputch was planned by Hitler inside the very hall you and I were just in. Hitler and about 600 of his followers stopped a rally already in progress at the Burgerbraukeller, a beer hall not far from here. Like some animal, Hitler is wild with excitement, so he leaps up onto the table and fires his pistol into the ceiling, announcing to the crowd that the national revolution has begun. In another hall just across town, at the Lowenbrau Keller, his accomplice Ernst Rohm is waiting with members of his own Fricorp, hoping to widen the coup. Like I said, their idea was for all of diss to spread to Berlin. But when Hitler fails to gain support from the leaders in his own beer hall, Rohm's men near the other one are easily defeated."[cii]

"I guess it's hard to believe it all actually happened. Everyone's been so friendly here. Countries change I guess."

"Yes. And Hitler vas wounded. They capture him and put him on trial where he undergoes this transformation. He vas put in prison but the judge liked him so he serves just nine months. Nine months. And unfortunately for Germany, he writes a book and became a well-known figure after dis.[113] So you see...a beer hall launched Hitler on his rise to power. It's not so strange, the yeast being involved in social change. Take your own history. You probably knew that the American Revolution was planned in taverns in and around Boston and New York didn't you? The Boston Tea Party? Or that John Adams and George Washington met for the first time in a tavern? Of course you knew. And how about Jefferson? Didn't Jefferson begin writing the Declaration of Independence inside a tavern? Of course he did. In less than ten years that document would change everything."

When we arrived at the Marienplatz I looked up at the stage, which is when it dawned on me I'd forgotten my eyeglasses back at the Hofbrauhaus. It was already dark and I was left with the light of the streetlamps to go by. I've needed glasses to see distance ever since doing too much reading back in graduate school. The scientific literature is a vast resource, or depending on how you look at it, a torture chamber, of small print and graphs all competing for the same space.

So as a result, when the bells of the Carolinian on the Glockenspiel began to chime and the life-sized wooden figures started circling high above our heads, all I could make out were some brightly painted, out-of-focus, blobs. I chalked losing my glasses up as punishment for drinking two liters of beer on an empty stomach while trying to keep up with a native Bavarian. I looked up at Wolf, who seemed transfixed by the display. It was as if he were seeing it for the first time. I took my pack off and placed it next to my leg, then dug into it looking for the guidebook so I could get an idea of what I was missing. The book had a photograph reminding me of a puppet show I once saw. The

---

[113] Berlin still has an "underground city" built in the 19th century by brewers to store beer since refrigeration wouldn't be invented until the end of that century. Hitler would later turn one brewing cave underneath Berlin into a weapons factory.

figures in the photo were all wearing traditional Bavarian costumes of the Middle Ages.

After about twenty minutes, the music was over and all the figures had disappeared back into their cloisters. Wolf didn't bother looking away during the entire show. I don't think he even knew I was there. And besides, it was too late to go back to the beer hall. My train was due to leave in just twenty minutes. On top of this, I still hadn't eaten or slept in two full days.

After the crowd dispersed, Wolf finally tilted his head down to look at me, then said softly, "You probably know by now I am a mycologist...a dedicated yeast person. I am...I mean I was a scientist and I don't think much of these gene jockeys these days."

"Yes, I think I figured that out Wolf."

"Bacteria and viruses with their DNA, they get all the headlines...and the grant money...and what's left for the rest of us? But that's how it goes, I suppose. The young take the place of the old. As it should be. Just the same, without fungi where would we be?"

"Up to our necks in undigested leaves?"

Wolf didn't smile, but instead continued speaking softly but deliberately. "You know, I cannot deny bacteria have their place. Did you understand the story?"

"Story?"

"The one the Glockenspiel just told us."

"No. I had no idea it was a story. Something about dancing?" I thought about my eyeglasses lying next to the pretzels but didn't mention it as I didn't want to give him yet another reason to look down on me.

"It was about the Black Death."

"But the figures on the stage...they were all dancing. Weren't they? The guidebook said they were." [114]

"Yes...they were dancing because as coopers of the town, the barrel-makers, they knew the plague was over. The year was 1517. Their job before was to make barrels for Munich's brewers, so they could store beer, and now the plague was gone and business was going to be good again, if only the people would come out and start living, you see? But the coopers, they were the only ones who realized diss. They promised the people that if they would come out of hiding, then they would do the dance for them in the Marienplatz. The people came out and the coopers kept their promise. In fact, you can see it. They still do diss dance every seven years...right here."

"So the people wouldn't even come out of their homes?"

"Some sealed themselves inside with bricks and mortar...they laid them in their doorways and windows. The wave of Black Death 150 years before, in

---

[114] The dance is the schafflertanz.

the fourteenth century, had already wiped out maybe a third of Europe...almost 40 million. Well anyway it kept returning, the rats and plague. No one saw the connection. So the people were afraid. It is quite natural. You are the expert when it comes to bacteria, perhaps you can enlighten me for a change." I couldn't tell whether he meant it as a compliment or something else. Then he cleared the air by adding, "You seem to know your history. Not many do these days. Except for Gamal, maybe."

"Well, the plague was caused by the bacterium called *Yersinia pestis*," I began, still wanting to impress him for some unsettling reason, "The disease had been known since antiquity, but not the microbe that caused it. It's even possible King Tut ascended the throne at such a young age because of plague."[ciii] I threw in some more ancient history as we walked because it seemed to calm him.

"It was bubonic plague that set the stage for the final events of the Roman Empire.[civ] It lasted 6 decades in Rome, killing millions. It had to have hastened Rome's downfall. Plague was also the final push that got the French out of Egypt in 1801. In the 1300's, ships from Asia carried black rats to Italy. Well when the rats died, the fleas feeding off them weren't too particular about where they got their next blood meal from, so of course they jumped onto the people, biting them and in the process transferring the microbes into their blood," I paused for a moment to make sure he was still listening.[115]

"Yes," he added calmly, still staring up at a full moon, which was just visible above the rooftops. He was clearly expecting me to continue. "Go on."

"An interesting thing about the plague is that it isn't even a human disease. Bubonic plague is a disease rodents get," I continued. "It's one of those zoonotic illnesses that happens when we get caught in the crossfire. In fact, about 75% of emerging diseases today jumped from animals...like the plague did...and still does. Maybe that's what's going on with Ebola now. The Ebola virus is becoming more humanized as it jumps from animals to us. Hopefully, Ebola and Hanta virus never evolves the tricks smallpox or *Yersinia* did...to cross over the human lung barrier and get spread through the air."

"Yes. I've heard many diseases came from animals. Because of agriculture," Wolf added.

"Lyme disease, measles, HIV," I rattled off a long list seared into my memory from cumulative exams back in graduate school, "smallpox, brucellosis, all came from animals. These are domesticated diseases that somehow crossed over. The flu virus is another...one that is still evolving

---

[115] Robert Hooke published a detailed drawing of a flea he made for his book *Micrographia* in 1665.

and jumping, evolving and jumping, becoming a little more human each time. We can't eradicate it, and probably never will. The virus remains safe inside birds and pigs. It rearranges its genes in pigs as if pigs were mixing vessels, and reemerges as a new strain of flu every year." [116]

"It was a big mistake killing all the rats," Wolf replied, steering the conversation back to something he clearly cared about, "People make so many mistakes. They thought the Great Pestilence was punishment from God. They practiced self-flagellation, whipping themselves on their backs. All they did was bleed so much that they gave the microbes to others. Their devotion made things worse, not better. People can be so silly," [cv] Wolf moaned wistfully.

"But at least there were some positive outcomes," I said. "Along with the horse collar, the coming of plague helped bring about an end to the feudal system in Europe, didn't it?[cvi] And because the people felt let down by the church they began doubting more. They had prayed but it didn't help. Ironically, it was because of the shortage of clergy due to plague that the colleges of Oxford and Cambridge (where the structure of DNA would someday be discovered, but I didn't mention that to Wolf) were founded, to train more priests. The loss of security in religion helped lay the groundwork for skepticism during the Renaissance. Modern science is based or skepticism, so in a way plague played an important part in bringing about the scientific revolution."[117] I started to wonder if taking me to the Glockenspie wasn't Wolf's way of apologizing for his harsh treatment of bacteriologists earlier on.

Wolf continued with my thought, "So the people found themselves in demand and were rewarded with a living wage for the first time. German farmers had disposable income. And do you know what they bought with their newfound wealth? You saw some earlier...in the beer hall we were just in. Mostly used as decorations these days. People can be so silly the way they relegate things to shelves."

It must have been obvious by my expression I didn't know.

"Would you like a hint? They were the first luxury items in Germany." I still drew a blank.

---

[116] The only animals the Native Americans deliberately bred were horses, which is why Lewis & Clark encountered so many Appaloosa after crossing the Divide and while living with the Nez Perce. The tribe routinely gelded inferior male horses. That day Lewis & Clark witnessed the largest herd of horses on the North American continent, tens of thousands while west of the Divide.

[117] Skepticism only *seems* new but there were actually ancient Egyptians that questioned (and even mocked) the belief of resurrection and an afterlife. Aristotle claimed that the mark of an educated mind was being able to entertain a thought without necessarily accepting it.

Embarrassed, I added, "No. I can't think of anything," wondering if he was disappointed in me. Strange how I still wanted to impress him for some reason.

"*Beer steins*, of course. They became more than just vessels to drink beer from. In time, beer steins found their way into the Bavarian hearts as works of art. They had scenes from the Bible or someone's family shield carved into them. A cottage industry emerged making pewter and earthenware beer steins, wooden and glass ones too. They found that, fired in a kiln with high temperatures, clay fuses like glass. It became less porous, stronger, and could even be shipped long distances. This new, harder surface was also more aseptic because it had less places for bacteria to hide out.[118] Germans thought drinking beer out of steins should be an event, a feast for the eyes as well as the stomach. But a strange thing happened along the way, you know? One of those little ironies of history we both seem to appreciate. Germany's new lager beers invented here in Bavaria were too pretty. Everyone wanted to look at their new beer...but not the container anymore. No, they only wanted to see through them," his voice became quieter, almost reverent, "And diss is why most people today drink beer out of a clear glass. It's a shame, really. No one cares about beer steins anymore. We changed the ale yeast into the lager yeast and look what happened. We Bavarians are good at shooting ourselves in the foot sometimes."

"But lager beers are more popular than ales. That's at least something, isn't it? When I think of a beer, I always picture a lager beer," I added in an effort to cheer him up for he seemed to be lapsing into melancholy again. "And what about the lids on beer steins? What's up with that? I've always wondered about the lids."

There was more silence and for a brief moment I thought that maybe he was ignoring me. Finally, he replied.

"Ah yes, the lids. Well they have a thumb lift and these also had to do with plague. The people were so afraid of plague returning. It was a constant fear for hundreds of years. Diss fear brought about the first hygiene laws for food. You see, in the late 1400's, while Columbus was discovering your part of the world, Germany was visited by swarms of flies, and since a connection had been noticed between filth and the plague, by the 1500's a law was put in place to have a lid on the beer steins...to keep the flies out." [119]
cvii

---

[118] This is also the reason your toilet and sink are made of ceramic. One of the reasons so little is known about ancient Greek houses today is that they were made of unbaked clay, which didn't stand up to the elements like fire-baked clay.
[119] Superstition was also why people killed the cats, which, it turns out, were partially immune to plague. A smaller cat population meant more rats and therefore more plague. / Biological warfare was waged in the Middle Ages when the bodies of plague victims were catapulted into walled cities under siege.

It was too bad he had to retire, I lamented. It was becoming clear that behind all the bitterness still lay a competent, perhaps even brilliant, mind. He was one of those people you sometimes come across in science, a maverick that you know belongs in research, and can also make a newcomer like me feel a bit inferior. Taking me to the Glockenspiel must have been Wolf's way of mending fences, I decided as we walked. I was slowly coming to realize this. And I'm not sure how much of it had to do with the beer, but I was definitely warming towards him. I decided as the train station came into view that he wasn't so strange after all. Just misunderstood more than anything. A genius, like Gamal had described him, but in his own peculiar way. I might feel the same way if I'd been forced out of academia the way Wolf had.

The next thing I knew, out of the clear blue, Wolf had me by my collar and with enormous strength began reeling me in like a fish.

"So tell me something before you go, dock-torr," he insisted as I arrived to within an inch or so of his face, "There is something I've been wondering about for a long time now...about your country. Are there any Schwarz in your department?"

"I'm sorry?" was all I could manage, close enough now that I could smell the beer on his breath and even see the gray whisker stubble on his Adam's apple, which seemed to be bulging to the point that it actually looked painful. I instinctively looked up into his eyes hoping for some signal he was only joking. Instead he merely cocked his head to one side, seeming more curious than anything, as if he hadn't done or said anything out of the ordinary.

"Schwarz...you know...black people," he reiterated and then paused, clearly waiting for an answer.

"Well...not in my department exactly," I squirmed, still firmly within his grasp, "but there is an African-American in another department." Before I could finish he was already nodding, as if he had all the information he required.

"This is what I thought. There are no Schwarz in Montana. They all prefer the big cities for some reason. You, on the other hand, you seem to like the mountains. Why is this?"

I didn't know how to answer him. It was true enough there were less Black people in Montana than most states for some reason. I never knew the reason. I never asked.

I craned my neck to look over my shoulder, realizing that the train was my best hope for escape.

"Well, I guess I'd better be getting off...if I want to find a seat. Haven't had a thing to eat all day," I added and began reaching towards the ground for my backpack.

He let go of me as suddenly as he had grabbed ahold, then hunched his upper body over into a ball as if contemplating something while I slung my bag over my shoulder. Diesel exhaust was already making its way into my lungs, replacing what was left of my appetite with a sickish feeling.

"It looks kind of crowded. Thanks again for everything."

"Yes. We should keep in touch, shouldn't we," he announced even though he'd never given me a way to contact him. Maybe he meant through Gamal.

YEAST MEETS WEST

After locating my seat, I relaxed and my appetite eventually returned. I was finally able to eat and even sleep for a couple hours. When I awoke, we were pulling into the airport at Frankfurt and a mad dash ensued thanks to my inability to read any of the signs without my eyeglasses on. With boarding pass in hand and only my backpack to worry about, I saved time and soon found myself on the right plane, doing my part to get ready for take off. Looking out the window at the baggage handlers placing luggage onto the conveyor belts, I couldn't help thinking about how much I'd learned. Just two short weeks ago I had little knowledge of the brewer's yeast beyond what it took to apply for a small protein expression grant. Now a fungus was taking the place formerly occupied since graduate school by a bacterium Robert Koch called *Bacillus anthracis*.

I also couldn't help thinking about Wolf, and how I'd known a couple professors back in graduate school like him. They were used to doing microbiology back when it was what other scientists condescendingly referred to as "stamp collecting".[120] What microbiologists did mostly for the first 100 years or so was go around finding new microbes, bringing them back to the lab and growing them on dishes and then characterizing them with various tests to determine what kind of food they required, whether they used oxygen to respire with or not, where they lived and so on, then they classified them by their appearance or ability to hold a gram stain. Things tended to fit neatly into nice little boxes in those days.

Then, something unexpected happened in the closing decades of the 20th century. DNA sequencing got cheaper and easier so that, by the turn of the new millennium, not only did we have the entire 3 billion base sequence of the human genome on an NIH computer, available to any person in the world with access to the Internet, but there were 20 complete genomes of bacteria

---

[120] Ernest Rutherford – discoverer of the atomic nucleus – famously said that all science is either physics or stamp collecting.

available too, base for base.[cviii]   And so because of DNA sequencing, things weren't fitting so neatly into little boxes anymore when it came to bacteria.   It was becoming clear that bacteria swapped a lot more genes with each other, even with viruses sometimes, than few had previously allowed themselves to imagine.   The insides of a cow turned out to resemble a trading bazaar not so unlike the Khan al-Kalili.   Swapping DNA is why bacteria can pick up antibiotic resistance genes so easily in hospitals.[121]   They get them from other microbes in our gut.

Instead of a "tree" with a trunk at the bottom and branches radiating outward towards the top, bacteria when they are grouped together on paper nowadays to show their interrelationships look more like an intricate cobweb, more like one of those food webs in biology textbooks used to show which animals eat which in ecosystems.   All in all, it can be a little disconcerting to see things you thought were once set in stone all of a sudden appearing to go their own way for someone used to a career based on absolutism.   It even calls into question the very definition of what a *species* is these days.   There can be so many shades of gray that some scientists think we shouldn't even be using the word species anymore when it comes to describing bacteria.   Things don't fit so neatly into boxes.   Some things are, for better or worse, a continuum the closer you look.   It's one of the reasons many of us in the life sciences still suffer from physics envy.   Yes, the days of "one gene, one Ph.D." were definitely over, I lamented as the earphones were being passed out and the trays placed into their upright positions.   The days of the amateur scientist like Leeuwenhoek working alone in his home making important discoveries were no doubt over as well.

Aviation science, while half a century younger than microbiology, is based on far simpler principles, and so within a few minutes we were safely off the ground and headed towards the Atlantic.   By the time the plane leveled out, the moon had slipped beneath the horizon and it was dark outside.   Everyone seemed intent on sleep but oddly enough I was getting my second wind and so I turned my overhead light on and took out the manuscript Gamal had entrusted me with, flipping through it, noticing that I was now about halfway through.

After the German doctor Robert Koch figured out a way to grow bacteria separately – like the ones that cause anthrax and cholera, on dishes with

---

[121] Bacteria mutate much more rapidly than our own cells.   In fact, it was mutant strains of anthrax that eventually led the FBI to their prime suspect in the letter attacks.   Dr. Bruce Ivins (the Ph.D. kind) did something unusual even among anthrax researchers.   Ivins had mixed mutant strains along with normal ("wild type") strains within the same flask of bacteria. This provided a unique signature (uncovered by directly sequencing the bacteria's DNA), allowing the anthrax used in the terrorist attacks to be traced back to his flask.   He committed suicide with an overdose of Tylenol in 2008.

nutritious growth media in the 1800's – it wasn't long before commercial brewers all over Europe and North America were using these same techniques to grow brewer's yeast in their own laboratories.   Being good businessmen, brewers simply wanted more reliable batches of beer.   It was more economical to control as many of the variables as possible.   Now, thanks to Leeuwenhoek, Pasteur, and the others, the yeast was one of those variables that could be controlled too.

Breweries like Carlsberg in Copenhagen, begun by the young Jacobsen, the one who had studied brewing in Munich, became pioneers of yeast research.   By the late 1880's not only were there methods for growing yeast as pure cultures, but brewery scientists were also able to keep track of the amount of sugar in beer wort using sacchrometers, and they could control the pH too, in fact the pH scale so familiar to us today was developed by chemists working at the Carlsberg Brewery.   The old hit-or-miss way of pitching yeast from previous batch into new one was, after thousands of years, finally on its way out.   It was too easy to pick up bacterial contamination that way.[122]   The time had come for a change.

For quite a while it seems, biologists classified the yeast as a plant, simply because it forms buds when it reproduces, looking the way a plant does when it forms a new bud on a stem.   In fact, it was this same budding process that finally proved Leeuwenhoek wrong and showed that the yeast was in fact a living creature when Cagniard-Latour took his turn examining it in 1835.   But it was Schwann who would come to realize that the single-celled brewer's yeast was – in truth – a fungus and so more like a mushroom, biologically, than any plant.   While working at the Carlsberg brewery in the 20th century, Dr. Ojvind Winge (the Ph.D. kind) looked closely enough to see that the yeast not only formed tiny buds, but could also undergo sexual reproduction with another yeast of the opposite type.[cix]

Two different brewer's yeast cells can, when conditions are right, actually fuse together to become a single cell so their chromosomes can mix. Genetic exchange occurs among their chromosomes to produce new combinations of genes, like shuffling a deck of cards to produce new hands during a poker game.   Biologically, the yeast would at this point be the equivalent to a fertilized human egg cell at conception.   But rather than developing into an embryo like we do, the yeast can live this way as a normal

---

[122] The Danish fermentation physiologist Emil Hansen at Carlsberg in 1883 became the first to work with brewer's yeast in pure culture.   He began his culture with a single yeast cell he isolated, the first time a pure culture was obtained after beginning with a single microorganism.   Similar to the way Pasteur distinguished bacteria from yeast in the sugar beet juice, Hansen was able to distinguish brewer's yeast from other species of yeast in nature. While many industrialists would have closely guarded their secrets for culturing yeast if they had discovered it, Jacobsen insisted Hansen always share his methods with other researchers by publishing in scientific journals.

yeast cell, except that now it has the ability whenever it needs to, of forming 4 hardy spores, or gametes, almost like going to seed. [123]   The closer brewing scientists looked at this curious phenomenon, the more they discovered.   Not only does the yeast cell have a choice of reproducing either non-sexually by budding, or by undergoing sexual reproduction to produce more diversity in its offspring, but it can even change its mating type, which would be like us going from a male to a female or vice versa whenever one needs to in order to increase one's chances of finding a mate. [cx] [124]

Having sex is nature's most efficient way of mixing up the DNA, of reshuffling the deck of cards that constitutes an organism's genome.   And since the Earth is an ever-changing place, life needs to change too in order to keep up.   This is what natural selection and evolution are all about.   Shuffling your DNA by creating new offspring is the best way to introduce more variety into your lineage.   Each new individual is, in effect, a new experiment nature is performing, and those best adapted to the changing conditions will survive and reproduce the best; the rest won't and so their genes will not get passed on, it's that simple.   This is, of course, the very definition of Charles Darwin's *natural selection*.   But something interesting happened in the yeast's association with humans.

Along the way, as a result of our hand in the process, the brewer's yeast became hobbled, sexually.   In fact, it's now recognized that most yeast used in brewing, baking, and winemaking these days can no longer have sex at all. After hundreds of years, brewers and bakers inadvertently and without even seeing the cells directly, selected for yeast strains that made the most consistent, reliable batches of beer, bread and wine.   So it was to their advantage to choose yeast strains that *didn't* change very much.   Therefore most yeast used in industry today can only reproduce by budding off identical pieces of themselves (i.e. clones).   They can no longer have sex and reshuffle their genomes anymore.   They lost that ability long ago thanks to mutations and our preference for them.[cxi]

Winemakers have even learned since Pasteur's time that the yeast contributes much more than just alcohol and carbon dioxide to wine.   In fact, it produces some 150 volatile chemicals – many aromatic – and the yeast even provides a valuable contribution to the "mouthfeel" and "cheesy flavor" of wines, which is why wineries will often leave their developing white wines in contact with dead yeast cells for several months after fermentation; during the aging process.

---

[123] For example, in nature if the yeast found itself in a situation where the juice of a berry was drying up and it needed to form spores to survive an impending drought.
[124] Scientists believe the brewer's yeast may have once been a multi-celled fungus, similar to the mushroom, but for unknown reasons reverted back to life as a single cell.

Wolf wasn't quite right about DNA being a dead molecule, though. One of the more useful aspects of DNA is that it can change, or mutate, its sequence. An important mechanism in evolution is, it turns out, dependent upon our cells making mistakes. When populations of cities began to increase during the Industrial Revolution, bakers could no longer keep up with the increased demand for bread by using the leftover skimmings from breweries. There just wasn't enough yeast to go around anymore. As a result, bakers began growing their own strains. Today, because of this, the baker's yeast has taken a turn and gone down a different path, evolutionarily, from its brewing cousins. Baker's yeast DNA these days have mutated in such a way as to allow it to become hardier, physically, so they can better withstand the rigors of modern manufacturing inherent in baking bread.[125] They are also able to produce carbon dioxide more rapidly (desirable for rising dough) and enzymes that are better adapted to consuming the sugars found naturally in bread dough. [cxii]

Today breweries like Carlsberg not only keep their pure yeast cultures separated in "yeast banks", but they may even use DNA fingerprinting, similar to what Tahany was doing to identify schistosome worms, but to keep better track of the changes in the yeast's DNA within breweries.[cxiii]

I flipped through the manuscript in my hands, cheating by looking towards the back to see how far I had to go. So far, except for losing my eyeglasses, the trip had been characterized by good timing. By the time the plane was over the Atlantic I had moved on to one of Gamal's later chapters entitled "Yeast and the New World". I looked back through what I'd read so far. Just a few circles I'd made, highlighting a typo here and there and some minor grammatical errors, that and a couple of suggestions in the margins about including some things I thought might be interesting, nothing to take away from the flow of it. Gamal had done his research...there was no doubt of that. Maybe a bit too much I was afraid. It tended to be dry in places – like the interviews with various mycologists I had never heard of – but all in all not bad. If it ended where I was now, I'd be glad to send it off to my cousin in New York as soon as I got back.

For some reason, Gamal began his next chapter with a discussion on distillation, for it seems that the only source of alcohol during much of western history was wine, mead, or beer, that is until around AD 800 when something new happened...or maybe it wasn't so new after all. Aristotle described the way seawater could be made into fresh drinkable water as far back as the 300's BC using distillation. But compared to distilling water, alcohol is a good deal trickier to obtain in purified form this way, which is why it took so long to

---

[125] Centrifugation, vacuuming, pressing into cakes, and freeze-drying of yeast. Before 1825 bakers used yeast from breweries in liquid form, or as a paste. In 1825 someone figured out a method to press yeast into a "cake", making it more easily measured out and sold.

attain pure, 100% alcohol. During the Middle Ages, Arab alchemists like Jabir had been making improvements in the still – an important one being the addition of what's called an *alembic*. This glass bulb extended the still's physical range, meaning it expanded the temperature gradient inside the still so that, when the alcohol became a gas and evaporated away from the heated wine down below, it could get farther away from the flame and so the vaporous alcohol could re-condense at the top of this cooler glass dome and fall like tiny raindrops, eventually coalescing and dripping down the sides of a tube leading outside the still, keeping the liquid separate from the heated wine inside, to be deposited as pure alcohol.

They called this new, combustible substance *aqua vitae*, or water of life, and it was so loaded with calories that Henry Ford would one day run his first automobile on the same stuff. The Moslem alchemists believed they had captured the spirit of the wine, but since the Koran forbids alcohol consumption, in the Moslem world pure alcohol remained mostly a curiosity, useful as a base for dissolving oils from plants to make perfumes, or for perhaps suspending various medicines in.[cxiv] Brandy, for example, is distilled alcohol obtained from wine, and the word itself comes from the Dutch meaning "burnt wine". And so it was up to non-Moslems like Arnold Bochmove in the late 1200's, to popularize distillation methods for making more diverse alcoholic beverages. His recipes caught on quickly in the west. [cxv] Being more concentrated, pure alcohol obtained by distillation was also easier to transport, not a small consideration in an age before steam travel was possible. Brandy, for example, is 8 times more concentrated than wine when it comes to alcohol content.

I leaned my head back into the groove of my headrest and closed my eyes, shielding them from the dry cabin and began thinking back to my American history class, the place I had first come across the importance of whiskey to the early settlers. It was something I knew Gamal could use in his book, if only I could remember all of it.

In a time before highways over the Appalachian Mountains, it would have been easier for farmers to transport their crop to market as a liquid, concentrated as corn whiskey. In fact, the only major revolt against the US government between the Revolutionary War and the Civil War was the Whiskey Rebellion of 1791, which began when the federal government tried to test its new authority by taxing whiskey to help pay down the Revolutionary War debt. The farmers of course thought the tax was unfair; being the very thing the colonists had fought against during the War for Independence.[cxvi] Meriwether Lewis took part as a volunteer. Even the president who ordered the uprising ended, George Washington, owned the most productive whiskey distillery in America, producing some 10,000 gallons of alcohol from grain in one year alone using 5 stills.[cxvii]

According to Gamal, within a hundred years after Columbus's last voyage to the New World, there were already so many vineyards in Mexico and South America that authorities back in Spain had asked King Phillip II to issue a royal edict discouraging the planting of more grapevines in the colonies. South Carolina had a vineyard as early as 1568 and Benjamin Franklin outlined winemaking principles in one of his pamphlets in the 1740's and within 13 years after the signing of the Declaration of Independence, there were perhaps 2,000 whiskey distilleries in the US. Washington and Jefferson were both interested in viticulture and improved on it whenever possible. I already knew that Jefferson considered himself a scientist above all his other trades, including politician, but was surprised to learn from Gamal that, back in 1620, the Pilgrims had stopped short of their intended destination, settling on Plymouth Rock mostly because the Mayflower's beer supply was running low.[cxviii]

From this point on, the chapter was a little difficult to discern and I wasn't sure why Gamal had arranged it the way he had, but I still managed to mine some interesting details before my eyestrain set in permanently the rest of the flight. I didn't realize, for instance, that it was the advent of faster clipper ships in the mid-1800's that allowed Bavarian brewmasters to bring their favorite lager yeast strains to America while the yeast were still alive, eventually expanding these yeast cultures in cities like Philadelphia, Milwaukee, and St Louis, or that the first paved streets in America were in New York City so the beer wagons from the nearby breweries didn't shake the kegs excessively, which could ruin the beer.[cxix]

I added a note in the margin recommending to Gamal he add a section on a subject I'm partial to…Lewis & Clark. Napoleon once said that an army should be supplied with just enough beer and wine to make it far enough so that no one would desert when the supplies ran out.[126] During the War for Independence, George Washington wrote letters to the Continental Congress when his soldier's alcohol ran low, so when Lewis & Clark set out to explore the newly acquired Louisiana Territory in 1804 looking for an all water route to the Pacific, it comes as little surprise they carried with them 120 gallons of whiskey.[127] They also held some in reserve to trade with the Native Americans. The party of 45 men and one dog set out from St Louis on May 14 and ran dry over a year later the following summer, near present-day Great Falls, Montana. And as Napoleon could have predicted, not a single soldier deserted when they ran dry.[128] The expedition did manage to obtain more

---

[126] He was also fond of bread sticks called grissini.
[127] Many of Custer's soldiers carried whiskey ("Dutch courage") in their canteens, in fact Custer's highest-ranking subordinate officer, Major Marcus Reno was, by several accounts, drunk on whiskey during the entire Battle of the Little Big Horn.
[128] The actual number in the expedition varied because some members joined while others left, which reminds me of a modern research laboratory in that way. / Early on, Lewis &

whiskey again before they got all the way back to civilization, on their way down the Missouri River, by trading information with settlers who were traveling the other way.

Gamal had somehow ferreted out some other details I'd never heard before, like the way Captain Cook learned from Native Americans how to prevent scurvy – by using spruce boughs – which Cook took along and fermented into a kind of beer on his voyages, and that alcohol was used in the first ships compasses to float magnetic needles in, and that the yeast's metabolism can also help reduce the natural toxins present in barley, which makes for a better tasting beer.[129]

Some of Harvard University's first students paid their tuition as malt grown on their family farms, and the benefactor of one of the earliest women's colleges in America was also a respected New York brewer named Matthew Vassar. He would later use half his fortune to start a women's college of the same name and brewers like Vassar relied on leftover yeast skimmings to help pay their bills. [130]

Gamal even had a paragraph on Prohibition for it seems that before Pasteur discovered heat pasteurization, there was no easy way to prevent grapes from fermenting into wine. Fermentation happens naturally because the yeast is always present on the grape so it gets pressed into service along with the juice. In 1869 an American dentist came up with what he called Dr. Welch's Unfermented Wine using heat pasteurization, which stopped the yeast from fermenting in its tracks, thereby allowing temperance Protestants to take communion using grape juice without partaking in the alcohol. His innovation also made it easier for Prohibition to be enacted in the 1920's.[cxx]

Just before drifting off to sleep somewhere over the Atlantic, I thought about Wolf one last time. I still didn't know what to make of him. That was the problem. Was he just being curious, or was there something more sinister to his question? And if so, was he aware of recent mitochondrial DNA evidence showing that the entire human race is more closely related to one another than a small band of chimpanzees are to each other in the wild? Growing up in the West there were some things I did know, though. I knew that just as Lewis & Clark had been the first Americans to reach Montana, so too was Clark's slave York the first black man to see it. History books hardly

---

Clark did court-martial two of their men for trying to steal whiskey – ordering 50 lashes on their bare backs. / When British soldiers occupied Monticello during the Revolutionary War, they helped themselves to Jefferson's wines.

[129] Two thirds of the Lewis & Clark Expedition suffered from scurvy, as meat became their main source of calories. None died of it, though. When the Pilgrims ran low and had to begin rationing beer on the Mayflower, scurvy struck shortly afterwards and two on board died (one was a sailor).

[130] Michigan's Lower Peninsula – which is shaped like a mitten – has a town in its *thumb region* named after Vassar. To this day the brewer's namesake is still straddling a portion of the Cass River that turned out to be an adequate site for a lumber mill in the 1840's.

ever mention that, or that they were boyhood friends and Clark had grown up with York because he inherited him.   In fact, York's father once belonged to Clark's father.   Slavery in those days was all in the family it seems.

SHUTTLE VECTORS

Within two days after being back, I found myself on yet another plane, this time headed west to Seattle, Washington.   It had completely slipped my mind – what with preparing for Egypt and switching to the new project – that I had signed up for a symposium about the yeast back when I was researching the grant proposal, which meant I was now making up for my lack of travel, all in the span of two weeks.[131]   Because classes were already starting for fall semester, I could afford only a couple days in Seattle, so I resolved to pack in as much as I could.   I wasn't thrilled with the prospect of taking another trip so soon, but if you want to know about horses, it's probably always best to go directly to the horse.[132]

While fumbling through the brochure at the convention center, one of the first things that caught my eye was a poster session.   It's a good thing the folks who arrange these get-togethers provide maps because the entire first floor was wide as a cornfield with row upon row of bulletin-boards, each with its own presenter standing next to their poster detailing some aspect of the yeast and eager to share his or her work with anyone showing the slightest interest.

The researchers who would draw the biggest crowds this day would likely be the ones who had published recently in journals.   Others would draw the curious simply because they were using the latest techniques someone might need.   These researchers had all the buzzwords in their titles, like shuttle vectors, microinjection, artificial chromosomes, and single-molecule detection. Poster sessions are a good chance to hear the nuts and bolts of research from the ones who are actually doing the experiments – on the front lines so-to-speak – often a graduate student, an undergraduate, or a postdoc with their mentor waiting silently in the wings to bail them out in case any of the questions get too technical (except for postdocs who often know more about the project than their mentors do).

---

[131] Unlike the Greek symposium with its 30 or so guests, and where Athenians could expect to drop by uninvited, I had to sign up for mine ahead of time and pay a fee.
[132] It wasn't until Champollion traveled to Egypt and saw the plants & animals of the region firsthand that he knew for certain hieroglyphics had to have originated in Egypt.   The Egyptians used what they saw around them (such as the Ibis) as inspiration for their hieroglyphs.

I was still in the process of filling out my nametag when I got wind later in the afternoon of a Nobel Prize winner scheduled to talk on the brewer's yeast, the same strain I was planning to use, and interestingly enough, he was a cancer researcher.   Seattle's Pike Place would have to wait as I set aside all other plans for the afternoon.   Most researchers drop everything they're doing at a symposium whenever a Nobel Prize winner is scheduled to talk. Attending a memorable seminar can be a mileage marker we gauge our careers by.

But try as I might, I just couldn't figure out how the brewer's yeast could shed new light on a disease as complex as cancer.   And, as I hope you know by now, the yeast is a single-celled fungus that spends its solitary existence in the soil or on vegetation and the like.   Yet cancer is a disease multi-cellular organisms like we humans get, a disease that produces tumors and spreads throughout the body.   Yeast don't get cancer because they can't form a tumor.   They're single-celled.

I looked at the map stapled to the inside of my booklet.   Whoever arranged the session had tried to group all the projects with similar themes together so as to avoid randomly zipping about the room.   Yeast projects in health and medicine were more towards the center, while projects having to do with industrial uses of the yeast were situated at either end of the hall. The farthest side was reserved for food scientists interested in making improvements in winemaking, brewing, and baking while at the end nearest me the yeast was being used to produce electricity in batteries, ethanol for cars, even bioremediation to help clean up toxic oil spills.   If I spent even a couple hours and made it halfway around, it should be time well spent.

So with this in mind I began nibbling away at the edge nearest me as if it were a pizza, strolling up first one aisle then down the next, ingesting, absorbing, assimilating, and then moving on to the next keyword or friendly face to catch my eye.   After attending a number of symposia, I tend to gravitate towards the less popular posters, where it's not as crowded and the students tend to look a bit unnerved, like a plucked chicken.   It's more relaxing and it makes them feel better, plus I get their undivided attention n return.   My first year as a graduate student I gave a poster presentation at this huge symposium and the only person who stopped by my poster the entire two hours was a competitor.

I had made the mistake of assuming my work would speak for itself, but scientists are people too and pay more attention to nice visuals just like anybody else.   My research was still unpublished and I also had the misfortune of being sandwiched between two other students on either side, both from Ivy League universities and had obviously spent a great deal of time and effort, not to mention money, on their posters.   Both veterans of previous campaigns they had the foresight to have them printed up professionally on one glossy sheet of laminated plastic in amazing computer-generated Technicolor like no poster I'd ever seen before. Just the simple experiments they'd done looked impressive, and maybe even more so was the fact that they needed only four tacks, one on each corner, to hold whole thing up, while I had been struggling with a whole box of tacks to display my meager shreds. As the morning wore on I began to feel more and more like a mud puddle the others had to figure out a way to jump across without looking too obvious in order to get on to the next beautiful poster. I thought back to what my grandfather liked to say about how life is always a lot easier when you plow around the stumps, except that when he said it, I had never counted on myself actually being one of the stumps. I couldn't wait for the whole episode to end so I could hide out among the crowd for the duration of the conference, like the insignificant researcher I felt certain I was to become.

I was slowly working my way around the perimeter when I came across a young man standing beside his poster and, judging by a lingering case of acne, was just now leaving behind the last vestiges of adolescence. You never know whom you'll run into at a poster session. It could be a Nobel Prize winner, a salesman trying to interest you in the latest piece of equipment you don't need, or it might be a gifted high school student taking extra classes at a nearby college. I noticed this one had braces on when he began explaining how his mentor was interested in using the brewer's yeast to generate electricity in fuel cells by feeding it organic waste from a landfill. He turned out to be a gifted high school student in an honors program, as I'd guessed.[133] His self-consciousness reminded me of myself at that age. He also had a good grasp of the history of the field and began by admitting how the idea of using microbes to generate electricity was nothing new, really. In fact, the same bacterium that lives inside all our intestines – *E. coli* – has been used to generate electricity in experimental batteries for decades, lighting dim flashlight bulbs in the corners of chemical engineering labs while being fed a continuous stream of glucose as an energy source, producing light in return.[134]

---

[133] It might surprise a lot of folks, but doing basic manipulation of DNA – like splicing a human gene into a bacterium – can be quite straightforward, in fact it's been compared to baking a cake. I've seen video of elementary students adding human genes to *E. coli* using the *heat-shock method*. The same *E. coli* that lives inside the human gut was the first organism genetically engineered by scientists in the 1970's.

[134] Glucose can let go of its electrons without the help of enzymes too. Ordinary chemicals will steal them, and in fact this is the basis for some readout devices diabetics use to monitor glucose levels in their blood.

THE YEAST AGE

Standing next to a poster a bit further down was the boy's opposite – a snappily dressed dark-haired city woman, probably an undergraduate with an eye towards medical school. She explained using her laser pen how her mentor had been putting their yeast to work detecting poisonous gas, presumably from a terrorist attack. Their goal was to float the brewer's yeast in clear plastic boxes in subways, not unlike smoke detectors, using the yeast basically as a "canary in the coal mine". The way it worked, she explained as her laser pen led me through all the intricacies of a rather impressive poster, was by making the yeast change color in the presence of the gas. She and her professor had genetically altered the yeast's DNA to produce a jellyfish protein that glows green whenever a poisonous molecule triggered a change in the output (i.e. "expression") of the "jellyfish gene".

Next down the line were two researchers engineering into their yeast some foreign genes from a grapevine. Their goal was to produce some of the same chemicals found in red wine that some think have an effect on preventing cancer and heart disease, but their project beer instead. Next up was an assistant professor who had engineered his "designer yeast" to produce aromatic chemicals, pleasing ones that made research labs smell better he promised, like bananas or strawberries; basically a yeast air-freshener fed a steady diet of sugar as its raw material.[135]

Another professor a bit further down was trying his hand at turning the brewer's yeast into a tiny vessel for making diesel fuel. Right across from him, someone had taken a gene from a fungus responsible for breaking down indigestible sugars like xylose (found as fiber in herbivore feces like the elephant's), and inserted this xylose-digesting enzyme's gene into his yeast n hopes the yeast could break down many of the unusable leftover sugars n plant bark and make more ethanol from it.[136] Another two researchers were planning on using brewer's yeast to detect the breakdown products of TNT basically by turning it into a bomb-sniffer. The senior of the two had gotten the detection gene from the nose cell of a rat and placed the rodent's receptor gene inside the yeast creating what they called a "biosensor". Not having gotten any new ideas, I soon had enough of industrial uses and moved on to the medical posters.

The pharmaceutical companies were all grouped into several aisles, some with a goal of making morphine in yeast, others focused on producing malaria

---

[135] One of the advantages of using brewer's yeast in the lab is that they tend to smell better than bacteria, reminiscent more of bread baking. Bacteria often produce more of a "rotting smell".

[136] Xylose is a type of glucose molecule chain that normally goes to waste when making gasohol.

drugs or fragments of monoclonal antibodies designed to target cancer.[137] Another had harnessed her yeast into producing toothpaste additives guaranteed to prevent plaque, and her colleague from the same company was after the Holy Grail of all yeast research...trying to engineer a strain that produces a chemical for his "Frankenwine" to counteract hangovers.

Finally, I came to the end of a long row where a student and her mentor were each taking turns in front of a large crowd like a tag team putting on a show, eager to explain how they'd gotten brewer's yeast to produce an enzyme that mopped up malic acid, the compound produced by grapes that makes wine taste sour.   What they were trying to do was a more sophisticated version of what the ancient Romans had already done: counteracting the sour taste in wine using marble dust, seawater, or ground up seashells, all of which are alkaline and so able to neutralize the grape's acidic (sour) taste.[138]

My confidence on the upswing, I ventured on towards the heart of the room where the intensity was turned up a notch and I would find researchers humanizing the yeast to produce proteins that cause Huntington's and other devastating diseases.   This allowed them to screen for potential life-saving drugs in hopes of "curing" the malfunctioning protein, the idea being that if you can correct a particular damaged human protein inside a yeast cell, then the drug has a better chance of working inside the person later on.   Another wanted to use his yeast to display molecules (proteins) on its surface belonging to the HIV virus that causes AIDS, with an eye towards using his yeast as a vaccine vehicle.   Whether he planned on putting the vaccine-coated yeast in bread or wine, I forgot to ask.   Another had inserted a gene suspected of causing Lou Gehrig's disease, or ALS, into the yeast and was testing chemicals (another drug screen project) to see if she could break up the clumps of aggregated protein associated with this disease.   Yet another had a yeast strain that produced the protein that causes mad cow's disease.

Eventually I found myself in front of a distinguished, older-looking gentleman with gray muttonchops as his only facial hair and an unlit pipe that had been extensively nibbled on.   Standing next to him was his student, clearly a first-timer who merely listened as the older man, an expert on progeria, explained in a serious tone how his lab was exploiting the fact that the yeast naturally has the same protein that causes premature aging in children, a genetic disease called Werner's syndrome.   When the mutated gene produces its faulty version of the protein, this damaged protein can no

---

[137] Malaria was well known in the Missouri River region in the 1800's, which is why Lewis & Clark took along 15 lbs of quinine-containing Peruvian bark as a powder.
[138] She got the gene to digest malic acid from a cousin of the brewer's yeast, *S. pombe*.   The ancient Romans had still another way to cut down on the acidity of wines: by tossing a piece of burnt bread into the wine jug, which is where we get the saying "to propose a toast" in English.

longer unwind the chromosome like it's supposed to during replication. Copying DNA is always a critical event that takes place just before a cell divides in two, for example when renewing tissue that has been damaged or simply gone past its prime.

A researcher right next to him was using the yeast as a window into normal human aging. It seems that when the brewer's yeast cell buds to form a new cell, it can do so only about 30 times before it begins to show signs of slowing down. This "elderly" yeast cell even becomes erratic, senile if you will, producing mutant yeast progeny with damaged DNA in the process. Understanding all the details of how this "mother cell" ages could yield clues for understanding human aging as well, he promised. I saw on his poster his grant had come from a pharmaceutical company so I questioned some of his data more thoroughly. One of the ways to increase the value of a drug is to find more patients who need the drug and so there is always this unsettling relationship between academia and industry, and sometimes it's a bit of a mystery trying to figure out what is real and what isn't. I became convinced after listening to him that he was honest.[139]

Some of the other memorable projects I came across that morning involved researchers genetically modifying the yeast so it could use synthetic amino acids not found in nature. The idea was to construct entirely new proteins with never before seen characteristics. Another planned on using her yeast to produce proteins from an arctic fish, natural antifreeze proteins which could then be purified and added to ice cream to preserve its smooth texture, while another was trying to create a yeast that could capture and convert carbon dioxide from the air into rock in the hopes of someday reducing global warming.

In fact, I became so engrossed in the last of these posters that I forgot all about the scheduled talk and as a result got there just as the first speaker had reached the wooden podium and was arranging his notes. My punishment was a seat in the very back of a theater-sized, echo chamber of an auditorium.

---

[139] The average lifespan of a yeast cell is about a week. / Another scientist was working on a joint project to sequence 1000 yeast genomes. They'd found so far that, rather than being domesticated just once, the brewer's yeast has done so multiple times throughout history in different places, and their chromosomes are mosaics that reflect this (when compared to brewer's yeast in the wild). It turns out there is considerable variation between different strains used in brewing today. Scientists have found that any two brewer's yeast strains, when compared to one another, can differ by 4 times as much as the DNA between a human and a chimpanzee differ.

NOBEL YEAST

As I sat waiting for the main speaker, something I had been hearing all morning kept coming back to me. When you hear the same thing over and over it begins to stick more. Everyone, it seemed, especially the ones just starting out like me, were having trouble getting research grants lately. Things were really tightening up over at NIH. Maybe I needed to be more resourceful, I reasoned. After all, before finding King Tut's tomb, Howard Carter had to sell watercolor paintings he drew for tourists just to make his ends meet. Even the ancient Egyptians built the pyramids without the benefit of a pulley since they didn't use wheels. But then again, who needs wheels when you live in the desert?[140]

The first speaker was a colleague of Hartwell's who, I could tell by the affection in his voice, must have known him for some time, and he set the tone for what was to follow, beginning by explaining how early biologists in the late 1800's had recognized cell division, or mitosis, when it occurred. They could watch this process single cells use to reproduce themselves unfold in surprising detail on account of how optics and lenses had improved considerably throughout the Industrial Revolution. In fact, by the 1880's their light microscopes were almost as powerful as the ones we use today are.[cxxi] Another advance, this one brought to the table by organic chemists, was the availability of newly synthesized dyes able to stain cells and their specific parts like the nucleus, parts that would otherwise remain invisible. Most of a cell is water and therefore transparent to light. Almost all parts of a cell need to be dyed to see them; which is how the word chromosome came about in the first place. Chromosome literally means "painted body" in Greek.

These early biologists, many of them German, knew that certain things had to happen when a cell divided in two, whether that cell belonged to an animal, a plant, or a microbe. Eukaryotic cells share many of the same processes in nature and the long thread-like chromosomes appeared under the microscope to these early investigators as if they were taking part in some kind of carefully arranged dance. A researcher in 1879 could have researched more than 200 scientific papers scattered throughout various journals describing chromosomes and the predictable motions they undertook inside cells as these cells divided into two. It was noticed, for example, that a cell's chromosomes always became visible and brightly stainable only at certain times throughout the process, and they tended to meet in the middle of

---

[140] The Romans invented concrete because – unlike the Greeks – they didn't have easy access to marble. An ingredient of concrete is calcium carbonate (limestone) and the Mars lander Phoenix analyzed Martian soil in 2008, finding it to be about 5% calcium carbonate. At least one astrobiologist has claimed you could grow asparagus in it on Mars.

the nucleus.[141]   They could see the chromosomes actually bending, as if being tugged at by some unseen force, pulled this way and that within the cell. The membrane barrier delineating the nucleus eventually disappeared, leaving the chromosomes in the middle of a large parent cell.

Then, like partners in the middle of a square-dance floor, the chromosome pairs each separated and went their own way, exactly halving in number, with a member of each pair moving away from the other to the opposite ends of the cell.   The final refrain came as the cytoplasm of the cell would simply pinch into two equal halves.[142]   Each new "daughter cell" ended up with exactly half the chromosomes of the "parent cell", no more, no less.   It was amazing and they didn't have any idea how it all happened.   Like Newton's laws that had been so successful in predicting the regular motions of the planets, or Galileo's when he discovered that the pendulum followed certain rules after noticing a chandelier swaying back and forth in a cathedral, it was as if the cell had laws of its own too, laws that were governed by an internal clock it could use to calibrate itself and keep track of what was happening during mitosis.

In the late 1800's, Dr. Theodor Boveri (the Ph.D. kind) had realized there were exceptions and that on occasion things didn't go so smoothly.   Boveri had been observing sea urchin embryos and worms when he noticed that, once in a while, one of the two daughter cells ended up with more chromosomes than it was supposed to have.   He even speculated that having this unusual number of chromosomes could explain diseases like cancer.   Boveri believed mitosis wasn't always perfect and that tumors could arise from these mistakes, errors that took place when one of the key actors got out of step during the "dance" that constitutes cell division.   Like the Earth's motion, cell division (mitosis) wasn't random.   But what was this unseen force, or glue, that held the whole thing together?[143] cxxii   No one seemed to know and few even bothered to guess for the next half a century.

Part of the problem was that animal models for learning about human diseases hadn't yielded much useful information.   In the 1930's a researcher could buy a strain of mice that would reliably develop cancer, and there were even certain chemicals – in coal tar for instance – that could be applied to the

---

[141] Brown named the *nucleus* in a paper he published in 1833.   The nucleus had been observed by others before, going all the way back to Leeuwenhoek, but Brown recognized it existed in all plant cells and he called it the "organ of reproduction".   He is the same scientist who discovered *Brownian motion* while examining pollen grains in water using a microscope in 1827.   Einstein would later use Brownian motion in 1905 to prove mathematically that atoms exist.

[142] They also knew different species had different numbers of chromosomes. / Even into the 1920's common belief held that cancer was caused by blunt trauma arising from a force strong enough to produce a bruise.

[143] The brewer's yeast has 16 chromosomes, but they're short and therefore couldn't be seen easily, even with high-powered light microscopes.   In fact, some biologists into the 1950's believed the brewer's yeast might not even have chromosomes.

bare skin of rabbits and guinea pigs to speed the process of forming a tumor along, but the mystery of what actually controlled cell division, and therefore cancer, remained out of reach. Even small laboratory animals like mice were too complex to get at these fundamental questions. Scientists needed a way around animal studies.

The first speaker had done his job and our appetites had been whetted. The stage was now set for Hartwell, a distinguished-looking, tall professor with a full head of grey hair and the eyes of a young man. He strode to the podium amid a polite applause. While arranging his notes I thought about what it must have taken to put his faith in the yeast while looking for answers to cancer. Here he was in the early 1960's, a newly minted Ph.D. interested in an organism so many had scoffed at when it came to complex animal diseases like cancer. He wanted to use a single-celled microbe as a model. On the other hand, maybe he was just doing whatever he had to, I thought. Researchers tend to be a resourceful lot.

Yet the scientific highway is littered with the careers of those who have gotten too overconfident, took a wrong turn early on, drifting too close to the edge of the scientific method in favor of their own version, creating a method so perverse as to not even be scientific anymore. The luckier ones figure it out early and get back on track while the unlucky never do, at least not until they've wasted decades, or even entire careers, not to mention all that money. "The easiest person to fool in science is yourself," as the physicist Richard Feynman was fond of saying, and King Tut's tomb may have contained a pair of blinders for his horses, but the truth is, blinders aren't just for animals. Scientists can wear them too sometimes.[cxxiii]

After a miscommunication with the person running the projector, Hartwell was soon under way, beginning by explaining how he came to be working with brewer's yeast in the first place.

There were many respected people who at the time believed the only way to understand what happened when a cell divided in two was to have all the pieces of the puzzle in place first...what amounted to a "master parts list" of all the genes and proteins involved. This is sometimes called the "bottom-up approach". The strategy is similar to compiling a program for a baseball game, where you have all the names and numbers identifying each player, his stats, and what each one's job is on the field. But Hartwell believed he could shed light on cancer by coming at the problem of mitosis "from the middle" so to speak.[144]

Thanks to Winge and the others over at the Carlsberg Brewery that had worked on the brewer's yeast for decades, it was possible by the 1960's to do some interesting experiments, mutagenesis experiments designed to damage

---

[144] Howard Carter came to Egyptology "from the middle". Lacking a university education in the subject, he was not as respected and therefore had to learn Egyptology as he went along.

only some of the yeast's several thousand genes, and then see what happens next. It's like removing a single part from an automobile at random – say the windshield wiper blade – and then seeing what happens to the car, or doesn't happen as the case may be. When the car doesn't drive as well in the rain, you might guess that the removed part had something to do with water and you might even speculate that the part was involved with removing water from the windshield. This is basically what gene-knockout experiments are a l about. They're designed to give you clues about a particular gene's function; leads you can later follow up on.

The brewer's yeast has about 6,000 genes (blueprints for how proteins are made), and these protein parts do just about everything a yeast cell needs to do, whether it's fermenting grape juice into wine, making bread dough rise, mating, or even copying its DNA before dividing into two new cells.[145] In the 1960's, Hartwell and his colleagues set about randomly knocking out each of the brewer's yeast genes with a chemical called a mutagen. If his hunch was correct, then it might be possible to identify key players in eukaryotic cell division by homing in on and identifying mutant yeast cells, hobbled ones that could no longer divide the right way, like Boveri's sea urchin cells so long ago that ended up with an unusual number of chromosomes. That was the idea at least, and it must have seemed a long shot at the time.[146]

But an important thing Hartwell had going for him was that microbes are so much easier to work with than animals. Still, most researchers in the field, leaders in cancer for example, thought of microbes as far too simple to yield useful information about a complex human disease. But the advantages were too tempting for Hartwell and this new generation of researchers coming up to ignore.[147] A yeast cell has perhaps five times less genes than a mouse cell, and yeast aren't picky eaters and so can be grown on Petri dishes as colonies. Each yeast colony is distinct and made up of clones (identical cells) and so each colony can have its own mutation in one particular gene.

On the surface of the agar in one dish it's possible to have as many as 300 different yeast mutants, each with a different gene knocked out of commission, while that many mutant mice would take up so much more space and require care and feeding, which of course adds up. Alternatively, a dozen or so Petri dishes can fit all stacked up neatly on top of one another on a shelf in some out of the way incubator. [148]

---

[145] By comparison, the parasitic bacterium *Mycoplasma genitalium* has just 517 genes, only 300 of which appear to be essential for life.

[146] Yeast has about 200 times less DNA than a human cell and can be grown in shakers and spread onto Petri dishes like ordinary bacteria. It's much easier than raising mice. And the animal rights people don't seem to mind, either.

[147] "Necessity is the mother of taking chances." Mark Twain.

[148] Some other advantages of microbes like the yeast are that they don't cause disease so they're safer to work with; they reproduce every hour and a half simply by budding, thus producing an exact clone (i.e. copy, or "genetic replica") which can then be grown into a

Hartwell wasn't too concerned about how similar or different yeast and humans were at the time. As he explained it that day, he was far more interested in finding specific genes inside a eukaryote cell involved in cell division. He was also gambling on the notion that even after a billion years of evolution that separated the brewer's yeast from humanity,[149] many of these same genes used for undergoing cell division were the same, still recognizable in other words, and that the yeast genes used for doing its cell division would be similar enough to shed light on our own reproductive genes. Rather than the painstaking bottom-up approach, as so many in the field were advocating, he was basically turning over rocks randomly to see what lay underneath; coming at the problem from the middle. Like my grandfather used to say, it's probably always a good idea to drink upstream from the herd if you can.[150]

Within days of mutating the yeast, Hartwell's team already had some results to follow up on. Many of the cells appeared under his microscope to be stuck in various stages of the budding process. Some yeast cells had tried to reproduce but had formed, like a bulge in the inner tube of a bicycle's tire that's about to pop, only a portion of a round new bud on its surface. And then that bud had malfunctioned and simply stopped growing for some reason. These new yeast buds (daughter cells) never detached from its mother cell like they were supposed to.

And what was perhaps most intriguing was the fact that there appeared to be specific steps during mitosis, and that these steps relied on previous steps in the process, almost as if the cell had quality control checkpoints along the way and these checkpoints could communicate with one another to find out what was going on globally inside the cell. Cell division was looking more and more like some kind of an assembly line, one that was hierarchical, and controlled by just a handful of key players. It seems there were more important genes and the proteins made from these genes were the foremen, or managers, and they went around deciding whether cell division could start, continue, and even stop if things went wrong. This would help account for the clock-like timing people had noticed for the past 100 years.

One mutant in particular called *cdc28* seemed to work at a critical point called the "Start Site" in the cell division cycle of the brewer's yeast.[151] Wouldn't it be interesting, Hartwell speculated, if the Start Site turned out to

---

separate colony (millions of identical cells) on a dish; they have simple nutritional needs including sugar, some nitrogen salts, and water, which is why people were able to cultivate them for thousands of years just by making bread, beer, and wine.

[149] The length of time since we last shared a common, single-celled ancestor.

[150] Now that I'm older and more well read, I believe my grandfather borrowed that one from Will Rogers.

[151] Which stood simply for *cell division cycle number 28*. Note that cdc28 is the name of the *gene*, while the *protein* made from the gene is called Cdk1. For the sake of simplicity, I'll use the term cdc28 to describe both the gene and the protein.

be the same place human cells had to cross when they too divided? And what if something went wrong there? Could it lead to cancer? [cxxiv] The finish line in a race is often the same place as the beginning. Perhaps cancer cells simply became deaf to the "stop" commands shouted at them near the Start Site, the same way a race car driver just keeps going around the track if he doesn't see the checkered flag. Maybe the cancer cell just keeps going through the cell division cycle, dividing in two endlessly even when they shouldn't...ignoring important signals. Cancer couldn't be that basic, Hartwell thought, but then again, maybe it could.

The next step was to find out what kind of protein the cdc28 gene made in the brewer's yeast cell and what it was that made this protein so important. It was more difficult to do this kind of work back in the 60's and 70's – perhaps years would go by with little to show – but eventually they did characterize it and the protein coded for by cdc28 turned out to be a member of a class of enzymes that had already been discovered, a class of proteins called *kinases*.

This was a ready-made clue for they could then speculate that the cdc28 protein was some sort of a messenger molecule that went around adding little chemical tags (called "phosphates") onto other proteins to get them to change their shape and thereby alter their function (adding phosphate tags is like adding tiny on & off switches to proteins). Enzyme reactions in cells are often cascade-like events, meaning that a change in just one single "upstream" protein like cdc28 can have dramatic effects on the rest of the proteins within that same assembly line but further downstream because a single cdc28 protein could, in theory, go around tagging so many other kinds of proteins with phosphates, which could, in turn, tag so many others, and so on.

When Hartwell published his findings on cdc28 in the journal *Science* in 1974, others including Dr. Paul Nurse (the PhD kind) in England, took an immediate interest. Inspired by Hartwell, in 1982 Nurse would go on to use a cousin of the brewer's yeast to show that the same cdc28-like protein existed in it too. Then Nurse helped make the remarkable discovery in 1987 that something similar existed in human cells, and that the cdc28 protein from the yeast could take the place of the damaged cdc28-like protein inside the human cell...step in for it in other words. [cxxv]

He found that the parts list for doing mitosis was interchangeable between cells of completely different species (like the way windshield wiper blades can sometimes be swapped between different makes of cars) even after a billion years or more of having gone their separate ways during evolution. They may not work perfectly, but close enough to restore normal functioning in the other cell. On a fundamental molecular level, our cells were looking more and more like the brewer's yeast than few had ever allowed themselves to imagine. [cxxvi] [cxxvii]

By the late 1980's thanks to Hartwell and the others, the details of mitosis were finally coming into focus, and it was possible to understand what it was that controlled the process that had so baffled the early cell biologists. Mitosis was a carefully controlled event. It has to be, especially in higher animals, so cancer won't result, and most remarkably, just a handful of genes controlled the whole process, and they did so in response to sometimes very different factors ("stimuli") in the cell's environment.

In the case of the brewer's yeast cell, cdc28 normally receives a signal from its surroundings to tell the yeast to divide when conditions are just right, for example when there is an abundance of sugars to be eaten. This stimulus, like the sound of a pistol beginning a race, triggers the cell to grow and then (with the help of enzymes) the chromosomes to form duplicates of one another.

But in the case of human cells, they normally pass their "start site" and begin growing and copying their DNA whenever they need to replace other cells in the body, for example when making new skin cells if a wound needs to be healed, the replacement of worn-out red blood cells, or the expansion of white blood cells to fight off an infection. Both humans and yeast were putting to work many of the same genes for controlling mitosis. Yeast cells turned out to be, in a surprisingly fundamental way, a very good stand-in for human cells because many of their genes turned out to be so similar. In 2001, Hartwell and Nurse were awarded the Nobel Prize in Medicine for using yeast to shed new light on human cancer.[152]

I sat in the back and waited through the entire question and answer period that always follows a stimulating talk like this one was. The whole thing was out of my area since I was trained as a bacteriologist. Bacteria have been separated from human cells three times longer than a yeast cell has and bacteria divide by a different process called "fission". They don't have their DNA inside a nucleus like yeast and human cells do so the whole cell division process is a lot different for bacteria.

But being simple is a good strategy and has worked well for bacteria. There are some that can reproduce in as little as 10 minutes...which is a mere 600 seconds! Human cells take hours. At one point, Hartwell did notice me, I think. Throughout the talk I had been sitting in the very back wearing a pair of sunglasses...the only prescription ones I owned at the time. I had left for Seattle so quickly that there wasn't enough time to replace the ones I'd left back at the Hofbrauhaus so my only other option was to wear the prescription sunglasses. What with my sporting a sunburn in a city famous for its lack of

---

[152] The third recipient, Tim Hunt, used sea urchin eggs. In the years since, it's been found that the brewer's yeast uses upwards of 800 different genes to accomplish mitosis.

sunshine, I couldn't help but wonder if perhaps Hartwell had assumed I was someone who had wandered in off the street high on drugs.[153] [154] [155]

Today, Hartwell's work with the yeast has influenced work in labs all around the world and it's now recognized that most human cancers are caused by mutations in the genes that control mitosis...defects in genes similar to cdc28. In fact, successful cancer drugs we now know often work by inhibiting this "runaway cell cycle" during mitosis and there are at least 23 different genes in normal human cells that can lead to cancer when damaged and perhaps most remarkably of all, each of these important, cancer-causing genes has a relative found deep inside a brewer's yeast. [cxxviii]

ALMOST HOME

Montana winters can be so long and cold that the hens have to lay standing up, as my grandmother used to say. I sometimes think I spend more time looking forward to summer, than actually enjoying it. After catching the redeye back to Missoula, I was as exhausted as I'd ever been working on the ranch and yet I still found my mind drifting back to the new project. When the ancient Greeks finished a symposium, they would parade through the streets of Athens singing together. I, on the other hand, had arrived home alone towards the end of an unexpected snowstorm. They seem to come out of nowhere this time of year, and this one brought with t the sense of dread that it was going to be a particularly long winter. It snowed on Lewis & Clark's men in 1805 as they traversed the Continental Divide not far from my family's ranch...and it was only September 3rd! [156]

Snow is the great equalizer, and accordingly a thick blanket of it lay on top of everything when I got back – cars, roads, houses, trees, and even lampposts – blending the living with the non-living into graceful ivory curves, the wind smoothing out all the sharp corners into diminutive drifts. It was all the tangible evidence I needed of what can happen when a dry Canadian co d front coming down from the north merges with a warm, moisture-laden Pacific one with just enough energy to make it up and over The Divide from the west.

---

[153] It turns out every human cancer cell has been found with a defect in at least one of ts genes that – like cdc28 – control the cell cycle and mitosis. / There are now new cancer drugs that target specific proteins in the cell that control division, as Hartwell had hoped.

[154] Unlike many wines, the yeast doesn't age very well. As a yeast cell gets older (they divide about every 90 minutes if conditions are favorable), the yeast even takes on some similarities to a cancer cell, for example its DNA chromosomes become more unstable, more easily mutated.

[155] Yeast uses the same 4-letter code in their DNA as our own cells use.

[156] It was so cold, there was no game and their salt pork was gone. Before getting to the western side, they ate some of their horses and tallow candles. Clark wrote "I have been wet and as cold in every part as I ever was in my life."

Thistle and the Lewis Valley were in its crosshairs. Seeing the snow reminded me of my grandfather explaining how the early settlers made their apple brandy – or homemade *applejack* as they called it – during the wintertime. In the autumn, after allowing the yeast to "harden" their cider into wine, they would leave it outside in the cold, whereupon the water would freeze solid, leaving behind just the alcohol as a liquid; a convenient way to concentrate alcohol and avoid the hassles of distillation.[cxxix]

Another hour of driving and I felt the warm sensation one often gets after they've passed the place they consider themselves home after a long journey. For me it was just after finishing the climb up to the top of the final ridge and before beginning my descent down into the valley, the point I usually considered myself home these days anyway. I thought back to the poster session while gazing across at the scene laid out before me in the soft, sparkling moonlight.

One of the researchers had mentioned that it was possible to buy through the mail a panel of yeast clones, each with a single one of its 6,000 or so genes knocked out ahead of time, including the same genes Hartwell had discovered. Another panel I could buy from the same company included each of the yeast genes tagged so that the proteins they produced would be easier to purify, or perhaps glow green, almost like each protein having its own bar code attached, the kind of barcode that makes keeping track of merchandise easier in a supermarket. We're so focused on knowing the structure of proteins these days, I lamented, stopping the truck alongside the road to take care of some personal business that couldn't wait. A yeast panel might be just the place to start when I got back.

Looking out over the guard railing at the rolling foothills beneath me, Thistle appeared as a miniature village, the kind you see under a Christmas tree behind a department store window. Barney was still at the ranch, asleep next to my mother's bed, no doubt protecting her in his mind, so I drove the rest of the way to the college, managing to find a space next to the microbiology building for a change. The entire lot was empty and the fact that the space belonged to the chair of the department made it that much better. Being a lowly, untenured professor often means parking the furthest away, like having the smallest office or drawing the least desirable committee assignments.

On the walk up to the building the lower cuffs of my trousers turned bright white as I blazed a trail through the new snow, but I hardly noticed as I was still too busy thinking about what would happen if I wasn't here in a year from now. Where would I be and who would be in my parking space? Losing out on tenure would be the first major defeat in my life and I had no backup plans yet.

I slid the snow away from the door with my boot, the sidewalk underneath still damp, as if the storm had taken even the ground by surprise. I pried open the heavy steel door and stopped just long enough to rid the powder from my boots by knocking my heels against the door, then walked carefully down the stairs so as not to disturb Charlie if he should be working on a night like this. There was some consolation in knowing that the ground was warm enough that the snow would all be gone by morning.

As I got closer to my office, I could just make out the outline of a package, a tall thin one in the dim light. It was leaning against my door and I as I got closer I could see it was a cylinder. I picked it up and placed it under my arm, fumbling around with my keys in the dark, then opened the door and turned on the lights, seeing now the colorful yellow and black stamps were in Arabic. I looked closer at the return address. It was from Tahany.

I pried open the metal lid with a key and slid the contents out. The paper gently unfurled, curling up at the bottom just before reaching the floor. I held the smooth papyrus up to the light. It was a replica of one of the scenes I'd come across in the museum that day with Gamal, a colorful scene depicting an ancient Egyptian family baking bread together. The scene was meant to insure the deceased pharaoh would have enough bread for the afterlife. They hadn't forgotten even though I had. Looking back, it was hard to imagine that just two short weeks ago I hadn't even met Tahany, or Gamal, or Wolf or even knew who Lee Hartwell was. Research can make for some unexpected acquaintances.

With the papyrus back inside its canister, I walked the short distance across the hall to my lab and began rummaging through my freezers and refrigerators, looking for what it was that had drawn me back this late at night. They had to be here, I reasoned, but where? There would be no sleep if I failed to see them again. But they weren't in any of the obvious places, so I checked the glass dehydration chamber, gently sliding the heavy, jelly-smeared lid over to one side to peer down through the jar's frosted glass but they weren't in there either. It wouldn't be until I had given up and sat down next to the computer in the back room, my head resting wearily in my palms, that I noticed them sitting on the bench right in front of me. The foil package of dried brewer's yeast I had turned to that night two weeks ago were right where I had left them.

I thought about how my search for the yeast had taken me halfway around the world and was just now bringing me home again, and about how Hartwell and the others have since gone on to show that over a quarter of our own genes have a counterpart deep inside the brewer's yeast. We're 25% alike, genetically, human cells and yeast.

I reached for a pair of tweezers, grasping some of the delicate crumbs, the prongs held lightly between my thumb and forefinger, managing to lift a few of the golden nuggets out and then I dropped them into a Dixie cup. Next, I added a teaspoon of water, agitating the mixture until it gradually disappeared...as if the crumbs had been consumed by the water. Then I transferred a small drop onto a glass slide, fixed a cover slip, and focused the fine-tuning knob on the microscope until I saw their faint outlines in the circle of light, as if a multitude of tiny actors caught in a bright spotlight. They were still there all right, round and intact, bobbing up and down in the water as I squinted. Each of the yeast cells a self-contained laboratory, floating right in front of me, each one a far better biochemist than I could ever hope of being, tenure or not.

I thought about how amazing it was that they had sat there in the same place night after night without any refrigeration, or nutrients, or even any water for that matter, as if waiting for me to come back and give them something better to do. The freeze-dried yeast cells hadn't changed in the slightest in the last two weeks, but I certainly had.[157] And it was also true that where I found myself at this particular point in my life wasn't the Mecca for scientific research every graduate student hopes of going to when they get their degree and go out into the world. And it was also true that I wouldn't be working on any glamorous diseases like anthrax, or TB, or the Bubonic Plague anytime soon.

But then I though about Tahany, and what she was accomplishing on her meager budget – of finding new and better ways of diagnosing schistosomiasis – and what a difference that would make to so many who didn't even know her, and probably never would. And then I thought about the yeast cells settling to the bottom of the water in front of me, and the opportunity I had been given to add to an already growing body of knowledge, of understand this amazing, versatile little fungus a little bit better. And as I leaned back in my chair, my feet propped up even with my body on the bench, I took a moment to savor the notion that this was the very first time since coming here as an assistant professor three years ago, the very first time that it all seemed like it might be enough for me now.

"God made yeast, and loves fermentation just as dearly as he loves vegetation."
- Emerson

#

---

[157] Since drinking beer or wine was safer than river water, modern humans have evolved to metabolize alcohol better than our pre-civilized ancestors could. In this way, the yeast is said to have altered the distribution of genes in the human population, allowing us to live in cities necessary for the Industrial Revolution. (i.e. a modern human could probably beat Neanderthal in a drinking contest)

The Secret Life of the Brewer's Yeast is part one of **Yeasts & Other Beasts: A Novel** If you'd like to read more about Dr. Ketchum and his adventures, this book can be found online at the same retailer.

~ ~ ~

Connect with me online at http://www.facebook.com/davidwoosterphd

~ ~ ~

SOURCES & ACKNOWLEGEMENTS

Among the many books and other resources that have been especially helpful are *Undaunted Courage: Meriwether Lewis, Thomas Jefferson, and the Opening of the American West*, by Stephen Ambrose; *A Field Guide to Bacteria* by Betty Dexter Dyer; *The Ancestor's Tale* by Richard Dawkins, *Innocents Abroad, Letters from Hawaii,* and *Life on the Mississippi,* by Mark Twain; *Anthrax: A History,* by Richard M. Swiderski; *Demon in the Freezer,* by Richard Preston; *The Last Stand: Custer, Sitting Bull, and the Battle of the Little Big Horn,* by Nathaniel Philbrick; *DNA: by Watson,* by James Watson; Dr. Leland H. Hartwell, Nobelprize.org; Sir Paul Nurse, Nobelprize.org; *Milestones in Microbiology* and *Robert Koch: A Life in Medicine and Bacteriology,* by Thomas D. Brock; *Pompeii: The Living City,* by Alex Butterworth and Ray Laurence; *Ambitious Brew: The Story of American Beer,* by Maureen Ogle; *History of Bread,* by Bernard Dupaigne; *Thermophilic Microorganisms and Life at High Temperatures,* by T. D. Brock; *A Short History of Nearly Everything,* by Bill Bryson; *Or Perish in the Attempt: The Hardship and Medicine of the Lewis and Clark Expedition,* by David J. Peck and Moria Ambrose; *Galileo's Daughter,* by Dava Sobel; *The Scientific Study of Mummies,* by Arthur C. Aufderheide, M.D.;

~ ~ ~

ENDNOTES

---

[i] Single cells staying together to become a multicellular organism was such a good idea, it happened throughout evolution several different times.

[ii] In graduate school I routinely used a special chamber with radioactive cobalt to *fry* anthrax spores whenever I finished an experiment, just to make sure they were not going anywhere.

[iii] Koch's mentor, Ferdinand Cohn, discovered in 1875 the spore stage as part of the lifecycle of anthrax.

[iv] Koch's work with anthrax eventually led to a method for growing pure cultures of other bacteria.   Before Koch's breakthrough, they were often grown as complex mixtures, which would lead to misidentification.   Some people employed elaborate explanations for having so many different shapes of microbe rather than accept the notion they might have contamination; for example, that they were looking at the same type of cell, but at different stages in its lifecycle while in reality, they were working with different species in the same dish (called *pleomorphism* this notion was championed by Carl von Nägeli).   The ability to work with pure cultures (a single type of microbe), as Koch pioneered, is therefore one of the most important milestones in the history of microbiology.

[v] Dr. Koch also went on to discover why anthrax could persist season after season: its spores were still viable even after drying out, sometimes for years.   Since Koch's day, spores of *Bacillus subtilis* – a relative of *Bacillus anthracis* – have flown to the moon on Apollo 16 and survived the vacuum and radiation of space for 22 months outside the International Space Station.

[vi] *Anthracis* probably forms spores as a way of competing with other microbes in the soil, like those that produce antibiotics for defense.   Chemical warfare is waged among unseen microbes all the time.   *Anthracis* "opts out" of this competition simply by outlasting whatever it finds itself with in the soil, by forming these protective spores and then going dormant as conditions become unfavorable.   Spore-making is so important to *anthracis's* lifestyle that it devotes one third of its genes to accomplishing this.   /   In 1997, researchers revived 80-year-old anthrax spores from a museum specimen.

[vii] A mixture of strong bleach & hydrogen peroxide will kill them on contact after about 2 minutes most of the time.

[viii] The smallpox virus uses a similar strategy.   And the spore-forming bacterium that causes tetanus, another close kin to *Bacillus anthracis*, causes lockjaw when allowed to invade deep within oxygen-less wounds.

[ix] Which would help explain how 94-year old Ottilie Lundgren was infected by just a few spores that rubbed off at a mail sorting facility and onto a letter bound for her home in Hamilton, NJ in 2001.

[x] In 2008 on Ted Turner's ranch near Bozeman, Montana anthrax spores that turned up naturally in the soil felled around 300 bison.   Fortunately, there hasn't been a human case of anthrax in Montana since 1961.   /   The Aum Shinr Kyo cult in Japan tried spraying anthrax spores from blowers at the top of an 8-story building in Tokyo in 1995.   They used a weaker strain of the bacillus, one for making vaccines.   Unfortunately, they then turned their attention to sarin gas with more lethal results.   /   *Anthracis* can, on occasion, become coated with silica without the help of scientists and achieve flight.   In Alberta, Canada during an unusually dry summer in 2000 at Wood Buffalo National Park, 42 buffalo kicked up sand to cool themselves and inhaled anthrax spores and swiftly died.   Contrary to news reports at the time, the anthrax used in the 2001 letter attacks wasn't weaponized.   The silica detected in the spores turned out to be a natural part of the spore

[xi] Like *Thiobacillus*.   While often vulnerable to changing conditions, bacteria as a whole are successful at inhabiting an astonishing variety of habitats on Earth, from deep-sea vents to freezing water, the atmosphere, and most everywhere in between.

[xii] It would no doubt increase it.   How much time a scientist spends doing actual research has probably always been important.   I sometimes wonder what the monk Gregor Mendel could have gone on to do in addition to discovering genetics if he hadn't become saddled with administrative duties later in his career at the abbey.   Galileo made sure he didn't have to teach at the University of Pisa before he took the position of professor. Theoreticians as well as experimentalists need their "alone time".   Einstein chose to work at Princeton over Caltech because Caltech wanted him to teach.

xiii Even Lewis & Clark were flat broke by the time they crossed the continent. They traded nearly everything they could do without to get to the Pacific and had to rely on their wits just to make it back. Lewis administered medical treatments to the Indians in exchange for horses to re-cross the Continental Divide. Their men found the Indians were fond of brass buttons on their jackets, so they cut off and traded them for roots to eat. The expedition carried with it a letter of credit from Jefferson, but had nowhere to spend it.

xiv The technology to *freeze-dry* microbes like the yeast is a 20th century innovation. During the Civil War, soldiers had to forgo the luxury of yeast when baking "quickbreads" around their campfires. To leaven bread dough, they relied instead on basic high school chemistry – a combination of baking soda & cream of tarter – which produces an acid-base chemical reaction along the lines of what happens when one drops certain tablets of antacids into a glass of water (plop, plop, fizz, fizz).

xv Flowers, tree bark, and other plant parts often secrete a steady stream of sugary fluids to the outside world as a way of encouraging friendly microbes like the brewer's yeast to take up residence there, space that might otherwise go to pathogenic (disease-causing) microbes (the *better the devil you know* strategy). Plants also make use of sugar as a reward for seed dispersal. Grapes contain sugars not for people to make wine with but for birds to spread their seeds. Some plant seeds won't even germinate unless they've first been through the digestive tract of a bird or an elephant.

xvi We have a planet in our solar system less dense than liquid water. If you could somehow place Saturn in an ocean large enough, unlike the yeast cell Saturn would float up on the surface. Rock here on Earth can attain a density less than water. During the early stages of the eruption of Mt. Vesuvius in AD 79, the pumice it ejected into the air landed and floated in Pompeii's fountains, which must have been a curious sight for the volcano's victims. / English King Henry VIII kept 15,000 gallons of wine in his cellar at Hampton Court and had fountains that flowed wine to impress his guests with an apparently endless supply.

xvii Actually, yeast do find a way to move – up and down in liquids while fermenting by changing their buoyancy – which is why anyone who brews ale will see a white foam develop on the surface towards the end of fermentation. Increasing its buoyancy is the yeast's way of escaping to the next batch, which in nature might be the next piece of rotting vegetation.

xviii Galileo, after building the best telescope in Europe, was offered what could be considered tenure today – a lifetime appointment at the University of Padua – in 1610. He turned it down to accept an offer by the Medici Family of Florence, becoming a professor at the University of Pisa. No fool, Galileo named the moons of Jupiter (which he discovered in 1609-10 with his improved telescope) the *Medician Stars:* to ingratiate himself with the Medici Family.

xix Karl Stetter, the German microbiologist who figured out a way to grow the majority of the world's extreme, heat-loving microbes in his lab (called the *witch's garden* due to elaborate equipment he had to invent just to culture them) is well known in microbiology circles for bringing his newly discovered extremophiles back to his lab...in an ordinary aluminum brief case. He also named the genus *Thermotoga* because the outer membrane of this thermophilic bacterium looks like a Roman toga. I also had a frozen sample of the thermophilic archaean *Thermoplasma acidophilum,* which has the admirable ability of growing & reproducing in either a burning coal pile or a hot spring, whichever it finds itself inhabiting.

xx A diary belonging to one of the astronauts from the space shuttle Columbia survived the craft's demise and even the 37-mile plunge through Earth's atmosphere, yet when it was discovered lying in a field in Texas 2 months later, its paper was undergoing digestion by soil microbes. Many of the space suits worn by the Apollo moon astronauts were stored in humid environments and as a result acquired black mold, a fungus that was consuming the polymer materials as food.

xxi 17th century chemists knew protein was unusual because fluids rich in them – like blood or egg whites – wouldn't boil but instead formed an opaque solid upon heating. Proteins must have seemed mysterious to medieval alchemists used to working with inorganic minerals and acids.

xxii It's these molecules that fall off bacteria during an infection that can get into the blood and cause fever.

xxiii Even *E. coli*, the bacterium that lives in our intestines, manages to stuff its DNA chromosome within itself in spite of the fact that, if fully extended, its chromosome would be 1000 times longer than the cell. The DNA inside one of your own cells would stretch 2 meters if fully unwound and joined end-to-end. Of course it would be so thin that you'd still need an electron microscope to see it. In fact, if you could stretch out and line up, end-to-end, all the chromosomes from all the cells in your body, your DNA would cross over 700 million miles; or to the Sun and back about 4 times.

xxiv The rise of the budget traveler can be attributed to a guidebook written in 1957 by Arthur Frommer called *Europe on $5 a day*.

xxv On their trip Twain and his fellow travelers formed a club that met onboard the *Quaker City*. Each member took turns reading aloud from guidebooks, learning about things they would see the next day. Even the ancient Egyptians were fond of travel stories. One was *The Shipwrecked Sailor*, a fantasy where a sailor returns to Egypt with exotic gifts having traveled to the Land of Punt.

xxvi The Art Deco movement in America was influenced by interest in Egyptian art in the 1920's while Napoleon's invasion of Egypt and the subsequent "Egyptomania" it inspired influenced women's fashion in Europe back in the 19th century. (He also took along a printing press to Egypt that had Arabic type) / The ancient Egyptians had no word for "art" or "artist". They saw the people who made wall paintings, sculptures, and other works as simply copiers. Egyptian art was never supposed to change, so creativity wasn't valued then like it is today.

xxvii An Egyptian would have started his or her day with some beer, yet did without money until the 6th century BC. Besides beer, payment might also have been in the form of sacks of grain. Stamped coins – what we'd recognize as money today – weren't used until around 350 BC, having been invented in Greece. Information was a 2-way street in the ancient world, which is how the Greeks got their brewing knowledge, in turn, from the Egyptians.

xxviii Just from 1450 to 1500, some two million books were printed in Europe thanks to Gutenberg's invention. Within a year after Galileo finished *The Starry Messenger* in 1610 for instance, there were already bootleg editions being printed throughout Europe without his permission. Before Gutenberg, books were a luxury item and a typical university student during the Middle Ages would have had to rent them.

xxix Anthrax is surprisingly common and found in soils throughout the world. Researchers comparing the DNA sequences of different strains have discovered the bacterium is actually several related strains still found along what was once an ancient trading route from China to Europe. The oldest strains were in China, intermediate strains towards the midway point, while younger strains evolved closer to Europe. The *Ames strain* used in the 2001 attacks is related to one found today in China, suggesting it arrived in the US via trade in furs sometime during the last 300 years and has now (for some unknown reason) settled in Texas and Oklahoma.

xxx One of the original 7 wonders of the ancient world, the Lighthouse of Alexandria, tumbled into the sea after earthquakes in the Middle Ages. But it could be claimed the landmark is still around. Just as the lighthouse had 3 stories, so too do many minarets of Cairo's mosques have them. So in a sense the lighthouse never really left. It simply became incorporated into the City Perpendicular's skyline.

xxxi Some strains can cause necrotizing fasciitis, or so-called flesh-eating disease. The 2001 Nobel Prize winning physicist Eric Cornell lost an arm and shoulder to this microbe. / One reason doorknobs are often made of copper is the innate antibacterial properties copper ions possess, some of which always come off onto your hands whenever you touch a copper doorknob.

xxxii The captain would have been given the copy while the library would have kept the original papyrus. By the time Julius Caesar made it to Egypt there were perhaps 700,000 manuscripts in the library. / The largest collection of ancient papyri ever uncovered (1,785 scrolls) was at the Villa of the Papyri in Herculaneum. Napoleon was given several of the charred papyri as a gift while Sir Humphrey Davy failed to unroll the damaged papyri using 19th century chemical techniques.

xxxiii Homer paid tribute to the skill of ancient Egyptian doctors in the *Iliad*, a work in which he also described anthrax as a "burning plague". Homer used the Greek word for bread ("broti") to mean "mortals" since only the gods could drink ambrosia while mortal men and women were stuck eating bread.

xxxiv Other languages have taken turns. While well versed in hieroglyphics, an Egyptian scribe would have written a letter to a foreign ruler in Babylonian – the *lingua franca* of the time. Later on (thanks to Alexander the Great), Greek would take the place of Babylonian as the most essential language of the Mediterranean. A merchant, mercenary, or someone attending the theater would have needed to understand Greek during Hellenistic times and is why many Romans had a Greek slave in their home: to teach their children the Greek alphabet. Before the first century, Judea was Hellenized, meaning Jesus would have known some Greek though his main language was Aramaic. Later still, Latin would predominate in the West (Newton wrote *Principia* in Latin, for example). When the poet Milton met Galileo in Italy the two were able converse because both men spoke Latin. Galileo was the first scientist to claim that the language nature uses is mathematics (subatomic particles today, for instance, can only be described using mathematical terms since physicists have few or no analogies in everyday life to describe them). Galileo wrote in Italian because he believed it important to reach as many readers as possible with his ideas and is also why his lectures were popular and the church considered him a threat. Galileo was unusual among scientists of his day because he also wrote about his discoveries in his native tongue.

[xxxv] There are many partnerships in nature, some extending into the microbial realm. Cholera is caused by the bacterium *Vibrio cholerae*, yet this diminutive comma-shaped microbe is completely harmless to humans until it first becomes infected by a specific virus, one that attacks only *Vibrio cholerae* (the bacteriophage is CTX-φ)

[xxxvi] The scientist declaring to know the exact species of a microbe while looking through a microscope is a myth invented by Hollywood. It usually takes several tests (for example, to determine what kind of metabolism it has), sometimes stretching over days, to identify with any certainty a particular strain of microbe. Does it grow on one type of media with sugar, or another with only acetate as a food source; does the microbe turn a particular dye blue, or does it leave the dye colorless; does it grow in the presence of oxygen or does oxygen actually inhibit its growth? Just as Koch did in the 1800's, a technician today might need to inject the pathogen into an experimental animal to see if it causes symptoms similar to the disease in humans before it can be said with certainty what the microbe is. Lager yeast and ale yeast can be distinguished in brewery labs by the temperatures they grow at and by their food preference. Ale yeast like a higher temperature while lager yeast strains prefer a lower temperature and can use the sugar mellobiose for energy. When spinal fluid was examined from the first victim of the anthrax attacks in 2001 it was recognized (using a microscope) as a type of *bacillus* by its box-shaped cells, but not more definitively identified as *Bacillus anthracis* until other more elaborate tests were performed at the CDC in Atlanta. An important clue that all the anthrax letters came from the same source was when the spores formed a mixture of different shaped colonies while growing on agar plates in the lab and this "signature" of different shaped colonies was the same from all the letters. It later turned out to be due to different mutations in the genes of the bacteria. The recent outbreak of cholera that killed 6,000 in Haiti was identified as being similar to a strain found in Nepal by using genome-sequencing techniques pioneered from the anthrax investigation of 2001.

[xxxvii] As detailed in the Hearst papyrus (so-named because William Randolph Hearst helped fund the expedition that discovered it).

[xxxviii] Species of mosquito that carry the dengue fever virus have evolved to live with humans in cities and have also changed their behavior, for example the times of day they take a blood meal: from early morning and evening (as most mosquitoes in the wild do) to during the day, when human activity is at its peak.

[xxxix] It is a phenomenon found throughout nature including the ancient world. Amplification was why the Egyptians had to be extremely careful when laying out the foundations for their pyramids. One small mistake at the pyramid's base could easily be amplified into structural instability towards the top. The 13-acre base of the Great Pyramid is so level that it barely changes in elevation by 2 inches.

[xl] Christianity is also why the Roman tradition of public bathing disappeared in Western Europe after the fall of the Roman Empire in the west. Christians being ashamed of their bodies avoided public nudity. Istanbul, Turkey – once the capital of the Byzantine Empire – is the only place where the tradition of public bathing has remained unbroken since ancient times.

[xli] Then again, "Faint heart never won the fair lady, neither did it ever pursue and overtake an Indian village" were some of the last words Custer wrote in a letter before arriving at the Little Big Horn. / The same week Custer's Last Stand happened, Mark Twain's *The Adventures of Tom Sawyer* went on sale in London (June, 1876).

[xlii] Cats were domesticated in Egypt around 2000 BC for the same reason pottery was invented: to keep stored grains safe from rats. So many cat mummies were discovered in

the AD 1800's that sailors used them as ballast for ships. Some cat mummies were then ground up and used as fertilizer.

[xliii] Hieroglyphs remained essentially unchanged even though they were used for 3,500 years and have a formal look to them because their original use was for monument inscriptions. It sometimes reminds me of the way the overall shapes of microbes like bacteria and certain fungi haven't changed either over millions of years...we know because many have been found locked inside amber 220 million years old.

[xliv] While searching for Tut's tomb, Howard Carter was reassured he was on the right track when fragments of wine jars thrown into a pit after Tut's funeral feast were identified nearby (his royal seal was on them).

[xlv] Clark saw a grizzly bear footprint before he or any American encountered a grizzly bear in the flesh. He described it as the biggest footprint he'd ever seen.

[xlvi] Acidic molecules are often negative in charge while clay tends to be positive so they bind each other. Many toxins in nature are negatively charged too, which is why elephants will eat clay from the bottom of watering holes. Clay absorbs plant toxins found naturally in the r diet. Some human travelers carry Kaopectate, which used to contain a type of clay with the ability to absorb toxins from bacteria consumed after a bad meal (now it has a similar substance called bismuth subsalicylate).

[xlvii] When Watson & Crick were determining the 3-D structure of DNA, they used data previously collected by Chargaff showing that DNA always came in equal amounts of the bases guanine (G) and cytosine (C) as well as adenine (A) and thymidine (T) (i.e. the amounts of G = C and A = T). To make his discovery, Chargaff used a strip of ordinary filter paper to separate the DNA bases since each base has a different "mobility" while traveling up the paper. Stanley Miller also used paper chromatography in the 1950's to identify the amino acids he created with an electric spark and some gases to recreate the earth's early atmosphere inside a sealed beaker. Paper chromatography can be used to separate (purify) many small molecules like chlorophyll and amino acids. In the undergraduate chemistry labs every year we have the students draw a dot of green ink at one end of a strip of paper, then dip that end in water and watch as the water carries the ink up the paper, separating the green into two new spots, one yellow, the other blue. Usually the blue one travels faster.

[xlviii] Fertile Crescent to the north of Egypt. In western Montana we get one crop a year while in ancient Egypt they got three thanks to warm weather and the generosity of the Nile. In ancient Pompeii, farmers often got two crops a year due to the unusual fertility of the local soils complements of Mt. Vesuvius.

[xlix] A method that can "weigh" molecules and thereby identify them by the specific atoms making up the molecule. The unknown sample molecules are first given an electric charge and then made to travel through an electric or magnetic field, which exerts a bending force. The heavier the molecule is, the less it wants to curve its path (due to inertia). The amount of curvature the molecule takes is directly related to the weight of the molecule.

[l] At some point during their evolution grapevines became self-fertilizing – or *hermaphroditic* – having both male and female parts on the same plant. Recent DNA sequencing of modern and wild grape vines suggests grapes were first domesticated in Anatolia (modern day Turkey) about 9,500 BC. / In a recent 2017 article, Andrew Graham and Patrick McGovern reported evidence of tartaric acid residue in 6000 year old pottery vessels in the Republic of Georgia. If true, this pushes the date of the first winemaking back almost 1000 years. Some of the Neolithic vessels had iconography including a cluster of grapes.

[li] The tomb of King Scorpion I held jars of resinated wine made around 3150 BC – about 4,000 liters worth when the tomb was stocked – as well as some beer and molds for making bread. The dried figs and grape skins, or lees (dregs), at the bottom of the wine jars were still in good enough condition to be analyzed. It too had become mummified due to the dry climate. Scientists used electron microscopes to see the original yeast cells, and even extracted some of its DNA. Wealthy ancient Romans were fond of wearing makeup and used dregs from wine to color their cheeks red. / The Incas of Peru made a beer called *chichi*, using saliva to digest the starch into simpler sugars, rather than performing malting as in the Near East (and in modern breweries today). In Honduras, the Indians were brewing a *chocolate beer* from the fruit of the cacao tree around 1000 BC. The Vikings brewed a barley beer they concocted out of bread soup, a source of much-needed calories during the winter. The Pilgrims carried barley seed to the New World to brew beer. / When the Corps of Discovery set out in 1804 from St. Louis, it was a mere outpost at the edge of civilization, a

small town lacking even a commercial brewery. When Clark canoed past the first settlement in Missouri – Ste. Genevieve, 50 miles below St. Louis – its brewery had been producing beer since about 1779. St. Louis wouldn't get its first commercial brewery until 1810. By the time Mark Twain was born in nearby Florida, Missouri in 1835 this small settlement already had 3 whiskey distilleries. / Before the American Revolution, cities in the east including Philadelphia had streets like *Brewer's Alley* set aside for beer makers. / Benjamin Franklin invented a recipe for spruce beer though he preferred drinking wine.

[lii] In 1866, Mark Twain spent 4 months in Hawaii as a correspondent for The Sacramento Union. In one letter to the newspaper he described the Kanakas (native Hawaiian agricultural workers) and their tradition of poi. "The poi looks like common flour paste, and is kept in large bowls formed of a species of gourd and capable of holding from one to three or four gallons. Poi is the chief article of food among the natives, and is prepared from the kalo or taro plant. The taro root looks like a thick, or, if you please, a corpulent sweet potato, in shape, but is of a light purple color when boiled. When boiled it answers as a passable substitute for bread. The buck Kanakas bake it under ground, then mash it up well with a heavy lava pestle, mix water with it until it becomes a paste, set it aside and let it ferment, and then it is poi – and a villainous mixture it is, almost tasteless before it ferments and too sour for a luxury afterward. But nothing in the world is more nutritious. When solely used, however, it produces acrid humors, a fact which sufficiently accounts for the blithe and humorous character of the Kanakas."

[liii] Fermented gruels that were porridge-like breads. The ancient Egyptians could turn ordinary bread into a "cake" for special occasions by adding eggs, fat, and honey to their dough before leavening. / It was likely the necessity to store grain centrally, and distribute it using the Nile as transportation that led to the formation of Egypt's Old Kingdom around 2686 BC.

[liv] One of the discoverers of oxygen – Joseph Priestly – lived near a brewery, which he visited often and is where he became interested in carbon dioxide (since this gas could extinguish candle flames yet paradoxically would allow plants to grow vigorously). The owner of the brewery let him set up a small lab, whereupon Priestly transferred water between two glasses over the tops of brewing vats to collect the carbon dioxide, in the process inventing the first soda water. He sold this recipe to a Mr. Schweppe. The new soda water sold well at health spas and made Priestly famous.

[lv] Moses and his followers fled Egypt so quickly they could take only unleavened bread, which is why unleavened bread is eaten today during Passover.

[lvi] It's interesting to consider the contrast in textures between the brown outer crust, which is crunchy, and the hot, moist interior of fresh baked bread, which is chewier. One of the reasons modern bakers use so many preservatives in bread is that bread today is made without as much crust, which would otherwise seal in moisture. Ancient bread in Egypt would have tasted more like sourdough bread today due to the presence of bacterial contamination. These stray microbes produce vinegar (an acid), which is sour. It's often incorrectly stated that yeast provide structure to dough by producing bubbles, because in actuality this gas weakens (tenderizes) the bread dough, making it easier to chew.

[lvii] Libations. There are several theories as to how civilization got started. Gamal also mentioned religion may have helped bring about humanity's first settlements. It's possible when people first began making regular pilgrimages to temples and other holy sites 10,000 years ago, the worshiper's requirement for food is what encouraged the first regular planting of crops.

[lviii] Memorizing ancient songs like the *Hymn to Ninkasi* would have been how illiterate brewers remembered all the ingredients and steps brewing beer calls for. In ancient Sumer there were 16 different types of beer, some made from barley, others wheat, or perhaps mixtures of the two. Brewing was done outside the temples in private residences – probably by women – and sold in nearby shops. Many ancient people later on, including Germanic tribes in Europe, had gods similar to Ninkasi, also devoted to brewing.

[lix] Egyptians brewed a wide variety of beers too including: "beer of the protector", "friend's beer", "beer which does not sour", "iron beer", "sweet beer", and "dark beer." Many were only for special occasions.

[lx] Memphis was the ancient capital of Egypt, not far from present-day Cairo. Its strategic location allowed for control of the Nile's Delta region, yet was still near the river enough to remain in contact with the religious center – Thebes – in Upper Egypt. Building materials from Memphis were later requisitioned to make Cairo in the Middle Ages.

lxi There are other stories like this that predate Christianity, for example: Imhotep, builder of the first pyramid in Egypt was believed to make barren women as well as fields fertile, and he was also half man, half god.   Another half-god was Egypt's Horus.   His father was Osiris, his mother the mortal Isis.

lxii The audience of perhaps 10,000 drank wine and ate meat while using bread as a napkin. Thespis was a member of the chorus – a group of 12 singers – and he developed a dialogue with the chorus, in the process becoming the world's first stage actor. / Unlike in Greece & Rome later on, ancient Egyptians had almost no public entertainment or spectacles.

lxiii Participants reclined around a circle (or square) so each would seem equal in importance. The Greeks also drank in rounds so all would reach the same level of intoxication at the same time. And yet the symposium wasn't a place to imbibe so much as it was a social situation where younger citizens were supposed to "display their minds", to show that they understood all the "rules" and that they were fit, upstanding members of Greek society. / The Roman equivalent to the symposium was the *convivium*. The same freedom the Greeks enjoyed to pursue knowledge (due to lack of a central authority) worked against them too, and is also why Greek city-states often fought amongst each other, eventually weakening their empire from within.

lxiv Temperature is involved in many interesting mechanisms in biology.   One example is the jacky dragon lizard. Temperature determines the sex of its offspring during incubation.   The temperature the embryos develop at inside the eggs can also influence the animal's behavior later in life.   Another example is yawning.   We probably do it to help cool the brain.

lxv There are some amoebae large enough to be seen with the naked eye. *Pelomyxa palustris*, for example, can grow up to half a centimeter across. / People have lost their lives to microbes in unexpected ways.   Methane gas is produced by a type of archaea called *methanogens* as a waste product, which has in turn exploded, killing sewer workers in the past.   Traces of methane have also been found in the Martian atmosphere, leading some scientists to speculate that it was created by a methanogen since methane is unstable to UV light and so something must be replenishing it. / The modern home has diverse ecosystems existing side-by-side, each with its own types of microbes.   Ovens are hot and dry, and many some of the same microbes that can be found in deserts like the Sahara, your freezer probably has some of the same organisms found growing above the Arctic Circle, salt can harbor species of bacteria also living in the Great Salt Lake in Utah. Even bacteria normally found in hot springs of Yellowstone National Park have been discovered living contentedly in hot water heaters. The International Space Station possesses the same exact species of microbes living on its walls as the average kitchen and bathroom does down on Earth. 55 different species of microbes had been carried into space by astronauts, subsequently shed, and were living on the walls. / After a dozen years in orbit, the Russian spacestation Mir was inspected in 1998.   The rubber seals around the windows were being eaten by colonies of bacteria, while electrical cabling was undergoing decomposition by another species of acid-secreting bacteria.

lxvi Robert Koch discovered the bacterium that causes TB in 1882. Even today it's estimated 1/3 the world's population is infected with the TB bacillus. It kills more every year than any bacterium on Earth, nearly two million.   Abraham & Mary Lincoln had only one child that lived to adulthood; two died of bacterial diseases.   Their youngest son Tad fatally contracted TB in 1872 while Willie died of typhoid fever in the White House during the Civil War. / Slave children prior to the Civil War were not auctioned off before the age of 8 because by then it was clear they had survived most serious childhood diseases.

lxvii The teeth of victims from Pompeii also show wear due to abrasive sand in their bread from grinding stones. Even bison in Yellowstone National Park can have this problem since grass growing near geysers contains silica due to the heat dissolving these minerals within the thermal waters and then being taken up by the plant's roots and stored in the leaves.

lxviii For many years it was a mystery to archeologists as to why the Greeks during the Early Bronze Age made wine vats with a hole in the bottom, since only the dregs would be found at the bottom of a fermenting vat of wine. Finally it was determined that the hole was needed to drain wine from the bottom because the Greeks also put a layer of olive oil on the top, which floated on the surface of the wine and served to keep air out during fermentation.

lxix As late as the 1890's wine was given as a medicine in Paris hospitals – some two million liters a year.   During Prohibition in the USA almost a million prescriptions for medicinal alcohol were written, one of ways people found to get around the law. / The first processed cheeses were made in the early 1900's by mixing wine into cheese. The wine functions as an emulsifier.

[lxx] Laudanum (or "tincture of opium") was carried by the Lewis & Clark Expedition – a mixture of opium and alcohol useful for alleviating severe pain and coughing among other ailments. The alchemist Paracelsus first made laudanum in the 16th century but its active ingredient, morphine, wasn't purified by a chemist until 1804. / After the death of her 3rd son, Mary Lincoln became addicted to laudanum, which may have contributed to her eventual mental decline. Laudanum was actually cheaper than alcohol in the mid-19th century.

[lxxi] Beer was also common in Rome. The most well-known beer lover in history, Julius Caesar, drank it after crossing the Rubicon. After the introduction of wooden barrels by the Gauls in the 1st century AD, the Romans made and stored beer in them.

[lxxii] Lead poisoning may have been why Roman aristocrats tended to have few children, a side effect being infertility. Julius Caesar fathered only one child (if that), by Cleopatra while in Egypt. / The higher lead concentration in the skeletons of men in Pompeii and Herculaneum is evidence that men drank more wine than women in the Roman world. Written on the entrance to a bar in Herculaneum: "...drinks cost an as here, but you get a better drink if you spend two. If you hand over four asses, you can drink fine Falernian wine."

[lxxiii] Ancient Rome tended to have more docile mobs, unlike Alexandria, which didn't have a dole, or "social sedative". This meant that Alexandrian aristocrats lived in fear of mob uprisings. Roman citizens eligible for free bread or flour included veterans, widows, and the unemployed. / A severe grain shortage in AD 6 led to nearly all the slaves being sent out of Rome and into the countryside to fend for themselves. The emperor Claudius had the port city of Ostia constructed to keep Rome supplied with grain from Egypt. / Ironically, Egypt today is the world's largest importer of wheat. Hosni Mubarak was forced from power in 2011 by riots fueled in part by poor grain harvests in Russia, which led to higher bread prices in Cairo. / In 2006, a US military patrol captured thumb drives belonging to al-Qaida detailing the organization's strategy for combating the surge. Among al-Qaida's targets were Baghdad's many bakeries. Their plan was to blow them up to deprive Iraqis of fresh baked bread in the morning.

[lxxiv] Some vineyards in Pompeii were recognizable to archeologists because Vesuvius erupted before winemaking season. Wine amphorae were still being stacked, which lends support to the August AD 79 date for the eruption rather than the November date as some others believe. The fertile soil around Vesuvius was so productive that locals believed the god Bacchus preferred the mountain as his home. Containers for ancient wine exported from Pompeii have been found as far away as the Red Sea.

[lxxv] Some proprietors watered down their wine much to the displeasure of at least one customer, as this nearly 2000 year old inscription found on one tavern's wall suggests: "Would that you pay for all your tricks, innkeeper. You sell us water and keep the good wine for yourself." Grapes were the most important crop grown around Vesuvius and artifacts depicting Dionysus are a common find in Pompeii, including mosaics, statuettes, busts, and paintings of him. Other graffiti on Pompeii's walls included political ads, with some candidates promising "good bread" if elected. / Even though it was a wine-producing area, wine amphora from Crete, the Aegean, and the Middle East (Gaza) have been uncovered in Pompeii. Apparently, Pompeii's wealthier residents valued the novelty of imported wines. / Lewis & Clark watered down their men's whiskey so it would last longer.

[lxxvi] Archeologists know from having uncovered an ancient "shopping list" that there were 3 types of bread in Pompeii: "bread" "coarse bread", and "bread for the slave". A daily staple, bread was the largest weekly expense in Pompeii. The average Roman spent 54 asses (13 sesterces) for bread, but only 23 asses for wine.

[lxxvii] Porters would have brought beer and bread to workers on site each day for lunch. For a worker in a temple or tomb in ancient Thebes, "brewing beer" or "having a hangover" was an acceptable excuse for missing a day's work.

[lxxviii] Typhus, typhoid, cholera & dysentery. The Crimean War is also where brucellosis was noticed for the first time...in horses. More recently, the presence of the bacterium that causes brucellosis has been detected in cheese recovered from Pompeii and the vertebra of several victims of the eruption also shows signs of brucellosis infection.

[lxxix] The idea of guidebooks isn't recent. The *Book of the Dead* was a series of papyri in ancient Egypt meant to help guide the newly deceased into the next world. Some were written on the mummy's bandages and described 200 magical spells.

[lxxx] Of all the animals on Earth, flying insects pose the greatest threat to humans. On a more benign note, the adult mayfly has the shortest lifespan of any animal…a mere 30 minutes. The fruit fly drosophila's generation time is just 10 days with the female laying 300 eggs each generation, meaning that two flies can produce 27 million offspring in a month, making drosophila an attractive choice for geneticists.

[lxxxi] The sprinklers irrigating the fields are a reminder that Egypt today stores more than just grain. Because of dams, Egypt also stores water in the form of large lakes not only for the water but also the hydroelectricity it generates to run the pumps for irrigation. Lake Nasser alone holds enough water to supply Egypt's needs for about 3 years.

[lxxxii] Cook was an English printer credited with inventing the modern tourist industry when it comes to package tours. As a temperance man, Cook viewed his excursions as a religious calling. Hotels his guests stayed in weren't allowed to serve alcohol and he printed the first informational brochures about sites along his tours. His devotion to the Bible is why he expanded his operations into the holy lands of Palestine and Egypt. Cook also saw an opportunity after the Suez Canal opened to give tourists a glimpse of Egypt. He realized that passengers would want to disembark in Cairo and stretch their legs. In the 1840 s (before steamboats were common) as in ancient Egypt four thousand years earlier, it would have taken about 20 days to travel by sailboat from Cairo to Thebes. No longer dependent upon the wind, by 1869 Cook's company was offering faster, more reliable steamboat rides up the Nile, cutting the same journey by two thirds the time. By the 1880's he had acquired a monopoly. In 1865 Cook was arranging tours of US Civil War battle sites for curious Europeans. By 1872 his company offered a 222-day round-the-world tour by steamship and stagecoach and introduced the forerunner of today's traveler's check in 1874. With his note a customer could obtain local currency in several countries. Cook invented the package tour, which included not only transport but also his guidebooks. Customers could expect food and lodging included in the price. In the late 1800's it was often said that the British had two empires in Egypt – a military one as well as a tourist one run by Thomas Cook.

[lxxxiii] Similar to the way Arab scholars and monasteries held onto classical knowledge during Europe's Dark Ages, ancient Egypt safeguarded Greek knowledge during Greece's own Dark Ages, which began around 1200 BC and lasted 300 years.

[lxxxiv] Even before the fall of Rome, religious travelers like St. Martin of Tours (AD 316-397) had gone on pilgrimages, planting vineyards and spreading viticulture. St. Martin visited the Touraine region of France and introduced the Chenin Blanc variety of grape. The ancient Romans expanded winemaking, creating some of the most productive vineyards in Western Europe today (in Spain, Germany, and France).

[lxxxv] The English word *yeast* can be traced to the Germanic word *gishen*, a term used by brewers to describe the roiling and foaming that accompanied a good fermentation. Today's beers usually have carbon dioxide or nitrogen gas bubbles injected after fermentation because the yeast has already been removed by centrifugation.

[lxxxvi] The Paulanerkeller was originated by Paulaner monks who migrated to Bavaria from Italy, while the Augustinians laid the foundation for the famous Augustiner brewery in Munich in AD 1328. Another nearby brewery is in a Benedictine monastery and may be the world's oldest brewery, founded in 1040. The Benedictines were founded by Saint Benedict, who believed monasteries should be self-supporting, which included the need for growing crops. One wonders if not for him, whether a later Benedictine monk would have been experimenting with pea plants in the 1800's.

[lxxxvii] The Domesday Book is the result of an extensive census compiled for William the Conqueror in AD 1086 for the purposes of taxation, and listed 42 vineyards throughout England. / One of the signers of the Magna Carta in AD 1215 was also a member of a winemaking guild. / The English word for "lord" originates from "loaf-giver".

[lxxxviii] The word *bridal* reflects this relationship, created by joining the words *bride* and *ale*. The word *honeymoon* may refer to the tradition of newlywed couples selling fermented honey to finance a trip. Another explanation has it that in ancient Babylonia the father of the bride was expected to supply the newly married couple with mead for one month.

[lxxxix] The ancient Romans put their stamp on Western cities even earlier; London is an example, founded around AD 50 and called Londinium. Six of London's city gates (including Newgate) date back to Roman times. / A European visitor to London during Shakespeare's time described English beer as being dark and cloudy, like horse urine. According to historian Barbra Tuchman in her book *The Distant Mirror*, Europeans in the 14th century had

no coffee, tea, tobacco, or potatoes, but "...hot spiced wine was the favorite drink of those who could afford it. The common people drank beer, ale, and cider."

[xc] As a rancher, I was surprised to learn that, aside from brewing, even today there are no additional uses for hops. The ancient Romans did find one, though – they put them in their salads – while Londoners in the 1660's burned hops in an attempt to ward off plague as they gave off such a strong smell. When early settlers to North America didn't have hops for brewing they often turned to spruce as a substitute. / Hops are a vine plant. Charles Darwin was fond of beer and did experiments with hops (the English county he lived in was famous for hops) to determine how the climbing vine was able to find its pole. Among his many achievements, Darwin was laying the groundwork for modern experimental botany today.

[xci] Anyone who wanted to become a baker in London circa 1300 would first have been apprenticed to a baker before being allowed to join a guild. / The first modern chemist – Robert Boyle – had an interest in fermentation and in the 1600's predicted that whoever came to understand the cause of fermentation would someday understand what caused disease. / The guild system of the Middle Ages, with its apprentice-mentor relationship, is what today's graduate schools in universities evolved out of and how I got my Ph.D.

[xcii] In 1789, a journalist in Paris reported: "The bread is black, gritty, and sour. Flour is of the vilest quality. Yellow in color and foul smelling. Clotting hard masses that could only be broken up by repeated blows of an axe." / The variety of French pastries today has its origins in government price controls of the 1780's. Because Parisian bakers could no longer charge as much for ordinary bread, they made pastries for wealthy clients containing luxury ingredients like sugar & chocolate, effectively allowing them to sell bread at higher prices. / A bread riot occurred in Richmond, Virginia during the Civil War in 1862 when hundreds of Confederate women took to the streets, breaking into stores in search of food and clothing. In 1917, bread riots in Russia would grow into the Russian revolution.

[xciii] Einstein tried and failed to get a job as a professor (or even a full time job as a high school teacher) after finishing college, due to his having rubbed influential professors the wrong way as a student. Einstein didn't have patience for those who held onto old views in physics without challenging them. Some have speculated that if Einstein had managed to land a job in academia sooner, he might never have come up with relativity. Working at the patent office gave him the freedom to pursue his own ideas. Isaac Newton also did the majority of his important work as a young man when he was forced to leave Cambridge in 1665 due to an outbreak of the Bubonic Plague. As Francis Crick once said, "by the time most scientists have reached age 30 they are trapped by their own expertise."

[xciv] Harvard University had its own brewery. / Beer brewed in monasteries is called *klosterbrau*. The world's oldest brewery – Weihenstephan – was built in 1040 and is near Munich. / Einstein was just 6 when his father's company provided the first electricity for Oktoberfest in 1885, lighting it up and refrigerating Munich's beer.

[xcv] It's estimated that at one time the entire human race probably numbered only about 10,000 - making us as endangered then as the great apes are today. / The beer hall I was in could hold 1,300 in the main room, the beer garden an additional 400, which is still a drop in the bucket compared to the Circus Maximus of ancient Rome, which could hold up to a quarter of a million. / A true beer culture, Germany's drinking age (for beer) is just 16. It's also not unusual to find labor contracts that stipulate the right of workers to drink beer during lunch.

[xcvi] Or "bottom-fermenting yeast". Rather than mule, most biologists would use the term "proto-lager" yeast. This lager yeast is also called *Saccharomyces pastorianus* (a hybrid of *S. cerevisiae and S. bayanus*) even though it was originally called *Saccharomyces carlsbergensis* by Hansen, who first described it in 1883. / A 2011 study by Diego et. al (published in PNAS) after searching around much of the world, identified *S. eubayanus* from Patagonia as the likely yeast that mated with the ale yeast *S. cervisiae* 500 years ago in Bavaria. / Bananas are a hybrid from the plant kingdom. Those tiny black spots are actually seeds that never formed completely because the modern banana is really the product of two different banana tree species that mated and produced sterile offspring. The only way banana trees can reproduce today is asexually (by producing "runners").

[xcvii] The word lager comes from the German word *lagern*: to store. Unlike ale yeast, which produce a sweeter beer and float to the top at the end of fermentation, lager yeast tend to sink to the bottom, and is perhaps how the hybrid was selected for: by staying in the bottom of the vats while the beer was poured out, therefore giving it a head start on reproducing when the fresh wort was put into the vat; another example of early brewers being

microbiologists and selecting for specific traits in a microbe, yet without ever knowing about microbes directly.   The colder temperatures of the ice caves and bierkellers also limit the number of side-reactions inside fermenting yeast, and helps to make for a clearer, cleaner-tasting beer appreciated in lagers.

xcviii DNA can add up. In the mucus of cystic fibrosis patients a common problem is too much chromosomal material accumulating in the mucus of the lungs.   This DNA is shed by immune cells.   One treatment involves inhaling an enzyme to break up this DNA, helping the mucus to flow more easily. Whenever we break open, or "lyse", a pellet containing millions of bacteria in the lab, so much DNA can spill out of the cells that the water turns to a gelatin that has to be repeatedly pipetted through a small hole to break up the DNA and turn the gel back to a liquid.   Proteins, on the other hand, can be both a blessing and a curse in brewing. Brewers usually try to get rid of leftover protein because it can cause cloudiness during cold storage. It's possible to add enzymes to chop this protein up (digest it) yet some proteins are needed to stabilize the bubbles in beer foam (foam being a desirable trait).

xcix I could have mentioned that it was these same "gene sequencers" he seemed to have so much contempt for who had recently determined the lager yeast was formed as a hybrid between two different yeast species hundreds of years ago in the bottom of a barrel in an ice cave (still others have sequenced ancient DNA of the yeast from the lees of Egyptian amphorae 4,000 years old).   A sure sign of a hybrid microbe like the lager yeast is when the cell has an unusual ratio of chromosomes in its nucleus.   This result of acquiring unusual numbers of chromosomes is called *alloploidy*.   The lager yeast has 60% more DNA in its nucleus than the more ancient ale yeast has in its nucleus.   Since they have been around much longer than lager yeast, ale yeast are, not surprisingly, the more genetically diverse.

c Long voyages in warm waters were always hazardous because of drinking water's tendency to become contaminated over time, which is why they added wine.   Convict ships to Australia, for example, had higher survival rates when they carried wine.   It's been theorized that water becoming undrinkable on ships is what held back discovery of the New World until the Renaissance.   Magellan's ship was the first to sail around the world, but it's often forgotten that he began with over 200 men and ended up with just 18 of the original.   Even Magellan died on the voyage (but not of bad water. He was killed by natives in the Philippines).   Pasteurization of milk is a way to kill the *brucellosis* bacterium that infects cattle, a bacterium that can also infect the consumer.

ci Before government agencies like NIH began funding basic research in the 20th century, science was done by wealthy "gentlemen scientists".   Scientific societies also awarded prizes for making discoveries. For example, the French Institute offered a one kilogram gold medal in 1779 to anyone who could explain fermentation.   By the end of the 19th century and until WWII, businesses would take the lead by investing in science, bringing about an end to the solitary scientist working in his lab. / Nationalism was alive and well in the 1800's. Pasteur, for example, was interested in beer not because he enjoyed it but so France could compete with Germany in brewing.   Pasteur made another significant contribution to microbiology, becoming the first to prove using his custom-made glass beakers that microbes were, after all, like visible (macroscopic) life and didn't arise out of nothing.   Microbes were everywhere, floating around in the air. He proved therefore that "even microbes have parents". Pasteur had put the final nail in the coffin of the *spontaneous generation theory*.   He even showed certain bacteria could, like the yeast, thrive without oxygen. / Details have always been important in science. When Pasteur was developing the first anthrax vaccine, he found he needed to subject the anthrax bacteria to temperatures between 42 and 43 degrees Celsius. This narrow band of temperature weakened the bacteria properly.   At other temperatures, they would form spores.

cii In *Mein Kamph*, Hitler described the first time he spoke to a large crowd (in the Hofbrauhaus; February/1921). He used a beer table as a podium. / In the beer hall pusch, Hitler took Bavaria's 3 highest officials at gunpoint into a back room to negotiate his terms while 3000 patrons in the front room grew increasingly restless. Fellow Nazi Hermann Goering tried calming the crowd by yelling "Shut up. You've got your beer, haven't you?" / When Hitler later ran for President of Germany in 1932, he promised the voters "brot und arbeit" (bread and work).

ciii The Italian doctor Girolamo Fracastoro (the MD kind) surmised that the Bubonic Plague might be carried by a kind of "seed" in 1546, a full one hundred years before Leeuwenhoek would observe bacteria for the first time. I've often wondered if Fracastoro hadn't seen a burr catching on someone's clothing while walking in the woods and that gave him the inspiration for his "germs". / In Paris as early as 1514, a brewing guild law forbade the presence of pigs, cows, and birds near brewing kettles because it was said to ruin the beer. / Pasteur's student – Yersinia – discovered the bacillus that causes plague, during the third pandemic.

[civ] In AD 542. Plague killed 10,000 a day at its peak in Constantinople between AD 541-2 (in 2014, DNA of *Y. pestis* was recovered from the tooth of a victim there). Large cities like London during the Middle Ages still had no underground sewage system, so waste was dumped right into the streets and the rats subsequently thrived there in great numbers. / Archeologists have never uncovered any communal dumps in Pompeii. It appears the ancient town's residents simply let refuse degrade naturally.

[cv] London's Great Fire of 1666 ended the plague outbreak in that city, but not until the disease killed 70,000. More than 80% of the city burned from the fire that began inside a bakery. This same outbreak of plague also closed Cambridge University, sending a young Isaac Newton home, where he would have what has since been called the most productive 20 months in the history of science. While away he laid the groundwork for calculus, the theory of gravitation, and began his work on light and optics. / In 2012, researchers traced several known strains of Y. pestis found in nature today (as well as the Y. pestis strain that caused the Black Death in the Middle Ages) back to a common ancestor strain that somehow diversified around AD 600 during the reign of Justinian.

[cvi] The horse collar was valuable because it allowed horses to breath while pulling heavy loads (by not cutting off their airway). This allowed horses to replace oxen for plowing and, since oxen plow 50% slower, a food surplus accumulated in most towns and villages.

[cvii] Plague usually begins as a tingling sensation all over the body. Before antibiotics were discovered, the unfortunate victim would likely have died a few days after symptoms appeared. In fact, plague victims were often still alive as their bodies began to decompose, some describing the smell of their own skin as it rotted, a result of the bacteria having traveled throughout the bloodstream and emerging on the limbs and torso. A common treatment for plague was bloodletting, which would have only increased spread of the disease. / In 14th century Munich as in many large cities of Europe, entire Jewish sections were purged when Jews became scapegoats for plague. Mark Twain once said, "History may not repeat itself, but it does rhyme a lot."

[cviii] Today there are over 700 complete microbial genomes sequenced and available at the NCBI website (NIH) to anyone with access to the Internet, in fact the entire DNA sequence of the anthrax strain in 2001 was available to researchers within a month after the first death in Florida. Wolf's concerns about protein chemists being relegated to obsoleteness weren't completely unfounded. A Holy Grail in genome research today is to someday use just the sequence of a gene to work forwards and determine the 3-dimensional structure of the protein it codes for. We already can get the sequence of amino acids of the protein simply by reading its gene sequence (looking at the DNA level), but knowing how the protein folds up into the unique and complex shape it has is an entirely different problem. Still, progress is being made and there is no doubt a Nobel Prize is waiting for the first person to do it reliably.

[cix] The prolific Winge – also known as the "Father of Yeast Genetics" – published more than 150 scientific articles, most on yeast, while at the Carlsberg Brewery. Carlsberg also helped pay for the study of young scientists, including Niels Bohr when he traveled to England to work with J.J. Thompson. In fact, the Niels Bohr Institute – which the brewery helped fund – eventually gave birth to quantum mechanics (Carl Jacobsen left his fortune to the Danish Academy of Sciences and Niels Bohr moved into Jacobsen's mansion).

[cx] This is a handy ability for a microorganism to have because microbes may find themselves needing to colonize a new location alone. In the case of the yeast, it would reproduce by budding. But if it couldn't change sex, all the offspring would be the same sex since budding only produces clones (i.e. copies). Fortunately for the yeast, it is able to change its sex. A popular saying among evolutionary biologists is that "sex increases randomness".

[cxi] The protein the yeast cell uses to detect the presence of another yeast of the opposite mating type is related to the same protein we have in our eyes called *rhodopsin*, which changes its 3-D structure in response to light (energetically, rhodopsin is like a mousetrap ready to be sprung). Rather than light, the yeast's protein receptor changes its shape in response to the presence of a hormone produced by a yeast of the opposite mating type, generating a signal that then triggers fusion of the two yeast cells.

cxii Damage to DNA often causes unwanted changes, which is why breweries keep their yeast strains in the dark, thus avoiding UV light that can damage (mutate) DNA. Brewers may also layer mineral oil over the top of yeast stocks growing on agar slants to keep out oxygen, or store the yeast in liquid nitrogen at -196° C, in suspended animation. / During WWII Fleishmann's Yeast (founded in 1868) developed a special strain of baker's yeast that didn't require refrigeration and could raise dough twice as fast as conventional strains, giving American GI's far from home the chance to eat fresh-baked bread, just by adding warm water to activate the yeast.

cxiii In Australia in 1923 the margarine-like spread *Vegemite* was added to grocer's shelves, having been invented by food technologists looking for uses for leftover yeast from breweries, the yeast being a rich source of B-vitamins and protein. Yeast used in brewing and baking are aneuploid, meaning they have extra copies of genes – backup versions so to speak – making it more difficult for them to change due to a single mutation (good for brewers but not so good for geneticists who like to change genes in order to study their effects).

cxiv Base is sometimes called the *mother liquor* by modern chemists, a throwback to the days when all chemistry was alchemy. Alchemy was invented in Alexandria, Egypt by the Greeks. Modern chemists still use several alchemical terms including alembic, crucible & alkaline. The word *alcohol* is Arabic. / Newton wrote one million words on alchemy, in fact it's been suggested he got his idea of gravity by pondering alchemy and its mysterious forces used to explain natural phenomena. / One of the reasons Arabs were the beneficiaries of advanced technology, like blowing glass and making lenses for example, was their close proximity to the ancient Library of Alexandria, the world's original "think tank". Arab scholars kept and transcribed texts that emerged from this melting pot that once existed in Egypt and where learned people from all over gathered for centuries to share, record, and develop new ideas. Cleopatra lived and was educated at the Library of Alexandria, which is why she could speak several different languages.

cxv As Gamal pointed out, due to the high alcohol content, brandy was an effective solution for dissolving herbs as well as honey, and is how Benedictine came about. The medicine was invented by Benedictine monks in Normandy. Alcohol also doesn't freeze as easily as water, making it a more convenient trade item for early settlers contending with harsh winter. Making brandy was also a way of putting to use poorer tasting wines that would otherwise have been discarded. Since alcohol boils at a lower temperature than water, brewers can take advantage of this when making so-called "near beer" (beer without alcohol in it) by simply making regular beer and then heating it to boil off the alcohol before bottling or canning.

cxvi They also used whiskey for barter and believed the tax unfair because it was levied on smaller distilleries more than larger ones. The government ended the rebellion but the tax was repealed in 1803. Distilleries located in Kentucky & Tennessee could make bourbon from corn without interference by the government during this period, which is why there are so many distilleries in those states today. The IRS was created in 1862 to enforce laws against making moonshine while taxes collected from liquor sales would help finance the Civil War. The first to *lobby* congress in Washington D.C. worked for brewers who of course wanted lower taxes on beer. In the early 1870's (before they were sent north to Dakota Territory and eventually fought under Custer at the Little Bighorn), the 7th Cavalry was stationed in southern states to enforce federal taxes on distilleries. The modern sport of NASCAR grew out of moonshine drivers testing their fastest vehicles against each other.

cxvii Unlike most whiskeys today, which are dark having been aged in barrels, Washington's whiskey would have been clear (since it wasn't aged). / Thomas Jefferson had two vineyards at Monticello and was known to experiment, growing different grape varieties. The third President also spent $7,500 on wine his first term in office (more than $100,000 in today's money). / Three decades before he became the 16th President, Abraham Lincoln held a state license to sell whiskey by the barrel out of his general store in Illinois. When Lincoln was just 5 years old in Kentucky, he used to carry lunch to his father Thomas, who worked part time in a local distillery. As for himself, Lincoln abstained from liquor all his life.

cxviii Each Pilgrim was allotted a gallon of beer per day during the voyage. The Mayflower was called a "sweet ship" because, before the Pilgrims chartered it, it carried wine & spices in the Mediterranean. The ship's hold gave off a pleasing aroma. The beer, bread, and other goods including tools, gunpowder and salted meats were kept in wooden barrels on the Mayflower. John Alden was therefore hired in England as the ship's cooper, charged with

157

repairing the barrels that would inevitably be damaged during the voyage. The Mayflower was classed as a 180 tun ship because it had the capacity to hold 180 barrels of wine (265 gallons per barrel). During the 1st Thanksgiving there likely wouldn't have been any pumpkin pie, but the pilgrims did have beer, brandy, gin, and wine. In fact, one of the first crops planted by the Pilgrims their first spring at Plymouth was barley so they could begin making beer again. A decade before the Pilgrim's arrival, the first ship to supply a colony in North America brought with it beer (in 1607). The Pilgrims also stocked the Mayflower with 20,000 "ship's biscuits" before leaving England. Ship's biscuits were a flatbread made from wheat flour and water, a key advantage being that they were more easily stacked. They also became infested with maggots by the end of the voyage and one pilgrim later recalled that the best flatbread for a trip of that sort would be so hard & dry that someone would need an axe to cut it. Once in the New World, the Pilgrims began making corn bread. Regiments stationed on the frontier during the Indian Wars (including the 7th Calvary) ate hardtack, a biscuit that could be up to 6 years old by the time they got it and sometimes required a hammer (or some beer) to break it apart. / North America does have native species of grapes, which is why the Viking explorer Leif Erickson called it Vinland ("wine-land") in AD 1000. These species (they may have been cranberries) tended to be more vigorous than Old World grapevines but sourer, as early colonists found out after trying to make wine from them.

[cxix] Wolf claimed back in the beer hall that the only real contribution America ever made to brewing was being the first to put beer in cans. According to him, American scientists developed a moldable plastic in 1932 that could line cans and thereby prevent beer from spoiling by keeping it out of contact with metal. At the time I thought he was joking. I'm still not sure. It is true that America was virgin territory for Bavaria's lager beers, not having had a previous tradition of drinking ales, like people in England had for centuries. So lagers caught on quickly in the USA. Milwaukee was also endowed with natural underground caves and the Miller Brewing Company has since turned one into a museum you can visit. I decided not to tell Wolf about one of my brothers winning a contest at the county fair for coming up with a way to deep-fry beer inside a pretzel. / St Louis has natural caves underground, which provided suitable conditions for lagering beer as was done in Bavaria (they used ice cut from the Mississippi to cool the caves throughout the summer). Some beer gardens in St Louis were located inside caves and by the 1850's one in three residents was a German immigrant. During the Civil War, soldiers smuggled guns in beer wagons and used the brewing caves as arsenals.

[cxx] Welch's innovation was the beginning of the processed fruit industry as well as Welch's Grape Juice. When Prohibition was first enacted, some towns sold their jails believing abstinence from alcohol would result in the end of crime. Binge drinking on college campuses today is a relic of Prohibition, back when youth crowded into speakeasies and drank quickly for fear of being discovered by authorities.

[cxxi] They had reached the limits of diffraction. The same developments in optics also took place with telescope lenses, paving the way for new planets and planetoids to be discovered. The scientist that discovered an optics law allowing for the correction of chromatic aberration in lenses was Joseph Jackson Lister, father of Joseph Lister. / Ground lenses have even been found at Pompeii, perhaps used by one of the town's engravers.

[cxxii] By the 1880's most trained physicians had moved beyond the view held by ancient Roman physicians like Galen and others throughout the Middle Ages – that tumors arose because of an internal imbalance of fluids – to believing that they were groups of normal body cells that had somehow turned "rebellious" and that this civil war could occur almost anywhere in the body. / In the 1880's, Wilhelm Roux suggested (correctly we know now) that the reason the cell division "dance" (mitosis) was so elaborate is so each new daughter cell can get the exact number of chromosomes it is supposed to have.

[cxxiii] Some early microscopists believed in the "theory of preformation" and convinced themselves they saw complete humans, called a *homunculi*, curled up inside sperm cells, complete with features like a head and torso. Others using telescopes in the 1870's convinced themselves there were elaborate systems of global irrigation canals built by extraterrestrials on Mars.

[cxxiv] Warts are tumor-like growths caused by viruses that carry special genes to cause the skin cells they infect to pass their "start site" and continue to divide when they shouldn't. Unlike most cancers, fortunately warts don't break up and spread to other organs. The host cell still

maintains some level of control. Calluses are interesting because they develop when cells under the skin pass their start site and divide again and again, driven by mechanical stimulation. They continue to grow when they otherwise wouldn't, thus producing a thickening of the skin useful for protection.

cxxv In "molecular biology-speak", Nurse *rescued* the mutant cell using the gene from a normal cell. Nurse rescued Hartwell's damaged *cdc28 yeast mutant* with the similar gene from his own yeast, a cousin of brewer's yeast called *S. pombe*, and it returned Hartwell's mutant brewer's yeast to dividing again like normal.

cxxvi cdc28 would turn out to be so important that this master regulator is regulated itself in at least three different ways, like a child that needs permission to go out and play from not only both parents, but must also be accompanied by a chaperone. In the case of cdc28, the chaperone that joins with it is a protein called a *cyclin*. Important processes in human cells are often strictly controlled like this. At each phase of the cell cycle, a different cyclin is produced, and each cyclin guides cdc28 to a new set of proteins, thus the reason for the perfectly choreographed dance of mitosis. There are also special "inhibitor proteins" the cell can make that will shut down cdc28. This is a remarkable level of control exerted over a single protein.

cxxvii As another analogy for cdc28: it's possible to picture it as the owner of a large department store, one who, before the store opens each morning, goes around telling each of his managers to begin giving out orders to their workers. The manager in charge of produce tells the workers in fruits and vegetables to start stocking grapefruit and other items found there, the manager of the finances tells his cashiers to start counting the money and loading blank register paper to get ready to take in the day's receipts, the manager in charge of maintaining the building tells his workers to start sweeping the parking lot and repairing any damage done the previous day, etc. cdc28 is like the owner of the store, and each of its managers would be the proteins that cdc28 binds to and modifies further down the line, for example enzymes on the surfaces of chromosomes which start copying each chromosome or line the chromosomes up in the middle of the nucleus afterwards, or break down the nuclear membrane at just the right moment so the chromosomes can separate and begin to form two new nuclei inside of the daughter cells, etc. Hartwell also found that the proteins acting later during mitosis can go back and put a halt to it further upstream if something were to go wrong. In the department store analogy, it's as if one of the store managers sees the cashiers have run out of register tape and tells the owner not to open the store until the situation is taken care of. Normal cells – rather than becoming cancerous when they're damaged – stop dividing in two and if this damage can't be fixed the cells will commit suicide as a safeguard against becoming cancerous. Biologists have a word for this cell death called *apoptosis*.

cxxviii Taxol, for example. / The first human cancer-causing gene (oncogene) scientists discovered – *ras* – is found in not only all normal human cells, but also the brewer's yeast. And *ras* in both yeast and human cells can replace one another because they do much the same job.

cxxix The seeds Johnny Appleseed distributed to farmers in Ohio, Illinois, and Indiana in the early 19th century were for apple trees useful for making cider only (he got the seeds from cider mills). / Several breweries in western states got their start in mining towns where German immigrants had given up searching for gold to brew beer near natural mountain springs. Some skipped mining altogether. Adolphus Coors stowed away on a ship to get to America and began gardening in Denver before starting a brewery in Golden, Colorado.